# 30th Workshop on Recent Developments in Computer Simulation Studies in Condensed Matter Physics 2017



Journal of Physics: Conference Series Volume 921

Athens, Georgia, USA
20 – 24 February 2017

ISBN: 978-1-5108-5341-6
ISSN: 1742-6588

**Printed from e-media with permission by:**

Curran Associates, Inc.
57 Morehouse Lane
Red Hook, NY 12571

**Some format issues inherent in the e-media version may also appear in this print version.**



Printed by Curran Associates, Inc. (2018)

For permission requests, please contact the Institute of Physics
at the address below.

Institute of Physics
Dirac House, Temple Back
Bristol BS1 6BE UK

Phone: 44 1 17 929 7481
Fax: 44 1 17 920 0979

techtracking@iop.org

**Additional copies of this publication are available from:**

Curran Associates, Inc.
57 Morehouse Lane
Red Hook, NY 12571 USA
Phone: 845-758-0400
Fax: 845-758-2633
Email: curran@proceedings.com
Web: www.proceedings.com

# Table of contents

## Volume 921

**30th Workshop on Recent Developments in Computer Simulation Studies in Condensed Matter Physics**

**20–24 February 2017, University of Georgia, USA**

**Accepted papers received: 30 October 2017**
**Published online: 19 November 2017**

### Preface



### Papers

## Preface

Thirty one years ago, because of the dramatic increase in the power and utility of computer simulations, The University of Georgia formed the first institutional unit devoted to the application of simulations in research and teaching: The Center for Simulational Physics. Then, as the international simulations community expanded further, we sensed the need for a meeting place for both experienced simulators and newcomers to discuss inventive algorithms and recent results in an environment that promoted lively discussion. As a consequence, the Center for Simulational Physics established an annual workshop series on Recent Developments in Computer Simulation Studies in Condensed Matter Physics. This year's highly interactive workshop was the $30^{th}$ in the series marking our efforts to promote high quality research in simulational physics. The continued interest shown by the scientific community amply demonstrates the useful purpose that these meetings have served.

The latest workshop was held at The University of Georgia from February 20-24, 2017 and served to mark three decades of successful and highly interactive Workshops. These Proceedings provide a "status report" on a number of important topics. This on-line "volume" is published with the goal of timely dissemination of the material to a wider audience.

These Proceedings contain both invited papers and contributed presentations on problems in both classical and quantum condensed matter physics. The day prior to the Workshop we co-hosted, together with Riken, Japan, a Tutorial on the simulations management tool OACIS, and the Proceedings begins with an overview of the tutorial material. The Workshop topics, as usual, ranged from hard and soft condensed matter to biologically inspired problems and purely methodological advances. While familiar topics like phase transitions were still on display, the trends in biophysics, dynamical behavior and complex systems demonstrated the continuing progression in the focus of computational condensed matter physics. We hope that readers will benefit from specialized results as well as profit from exposure to new algorithms, methods of analysis, and conceptual developments.

Athens, GA, U.S.A.
May, 2017

*D. P. Landau*
*M. Bachmann*
*S. P. Lewis*
*H.-B. Schüttler*

# Peer review statement

All papers published in this volume of *Journal of Physics: Conference Series* have been peer reviewed through processes administered by the proceedings Editors. Reviews were conducted by expert referees to the professional and scientific standards expected of a proceedings journal published by IOP Publishing.

IOP Conf. Series: Journal of Physics: Conf. Series **921** (2017) 012001 doi:10.1088/1742-6596/921/1/012001

# An open-source job management framework for parameter-space exploration: OACIS

**Y. Murase[1], T. Uchitane[2] and N. Ito[1,3]**

[1]RIKEN Advanced Institute for Computational Science, Kobe, Hyogo 650-0047, Japan
[2]Research Institute for Economics and Business Administration, Kobe University, Kobe, Hyogo 657-8501, Japan
[3]Department of Applied Physics, Graduate School of Engineering, The University of Tokyo, Bunkyo-ku, Tokyo 113-8656, Japan

E-mail: [1]`yohsuke.murase@gmail.com`, [2]`uchitane@rieb.kobe.ac.jp`

**Abstract.** We present an open-source software framework for parameter-space exporation, named OACIS, which is useful to manage vast amount of simulation jobs and results in a systematic way. Recent development of high-performance computers enabled us to explore parameter spaces comprehensively, however, in such cases, manual management of the workflow is practically impossible. OACIS is developed aiming at reducing the cost of these repetitive tasks when conducting simulations by automating job submissions and data management. In this article, an overview of OACIS as well as a getting started guide are presented.

## 1. Introduction

In this paper we present OACIS (Organizing Assistant for Comprehensive and Interactive Simulations) [1], an open-source job management software for large-scale simulations (`http://github.com/crest-cassia/oacis`). When we conduct numerical simulations in scientific research, we usually carry out many simulation jobs changing models and parameters by trial and error. This kind of trial-and-error approach often causes a problem of job management since a large amount of jobs are often created while we are exploring the parameter space. For instance, submitting one simulation job would require the following tasks.

- login to a remote host
- make a directory for a job
- create a shell script to submit a job
- submit your job to the job-scheduler
- wait until the job finishes
- transfer the result files to the local machine
- conduct statistical analysis on the results
- make plots

This process is repeated until he or she gets satisfactory results. Such repetitions are not only troublesome and tedious but prone to human errors. In order to conduct research in an efficient, reliable, and reproducible way, a software to help management of simulation results are strongly expected.

In case of social simulation, the problem can be even more critical because the amount of parameter space exploration is often larger than those for physical or chemical phenomena [2]. One of the significant difficulty in social simulation is the lack of established model. The models for physical simulation are usually based on solid first principle equations, such as Schrödinger equation or Navier-Stokes equation, which are carefully verified by real experiments. On the other hand, the models for social simulation are inherently phenomenological ones. These models are often constructed based on the observation of real society and/or our common senses. Even though an increasing amount of data about society is available, these data are inherently incomplete due to technical and privacy reasons. Therefore social simulations are inevitably less reliable compared to conventional physical simulations hence a single simulation run is just one of the possible futures. To obtain robust and universal information which is independent of minute details of model assumptions, we need to investigate various models and understand the global landscape of these models, which requires us to explore vast parameter spaces.

Using OACIS, we can automate most of the above tasks, which lets you efficiently explore the parameter space. With a user-friendly interface, you can easily submit various jobs to appropriate remote hosts. After these jobs are finished, all the result files are automatically downloaded from the remote hosts and stored in a traceable way together with logs of the date, host, and elapsed time of the jobs. You can easily find the status of the jobs or results files from a web browser, which lets you focus on more productive and essential parts of your research activities.

The development of OACIS has started in April of 2013 and is released as an open-source software in 2014 under the MIT license. After the first release, minor version-upgrades have been iteratively released which includes various improvements and bug fixes. In Oct. 2015, a major version upgrade, version 2.0, was released which includes significant enhancements. It has been applied to various research projects including studies on network science [3, 4, 5], molecular dynamics simulations[6], and agent simulations [7] to name a few. Functions needed by these projects have been implemented and usability are validated in actual projects.

## 2. Overview

### *2.1. Data Structure*

The system architecture of OACIS is depicted in Fig. 1. The web server is developed based on the Ruby on Rails framework (`http://rubyonrails.org/`), which provides an interactive user-interface. MongoDB (`http://www.mongodb.org/`), a document-based database, is used as data storage. The application server is responsible for handling requests from users. When a user creates a job using a web browser, the record of the job is created in the database. Another daemon process, which we call a "worker", periodically checks whether a job is ready to be submitted to a remote host. If a job is found, the worker generates a shell script to execute a job and submits it to the job scheduler on the remote host via SSH connection. Hereafter, we call these remote hosts "computational hosts". Then the worker process periodically checks the status of the submitted jobs. When one of the submitted jobs is finished, the worker downloads the results and stores them into the file storage and the database in an organized way. Hence, users do not have to check the status of the remote hosts by themselves and they can trace the simulation results even after several months. Various logs about the jobs, including the values of parameters, executed dates, elapsed times, are automatically kept by OACIS and they are stored in the database. On the other hand, the files generated from the simulators are stored on a file system, which enables you to directly access the simulation outputs using a file browser or a command line terminal.

OACIS stores the simulation results in a three-layered structure ("Simulator", "Parameter-

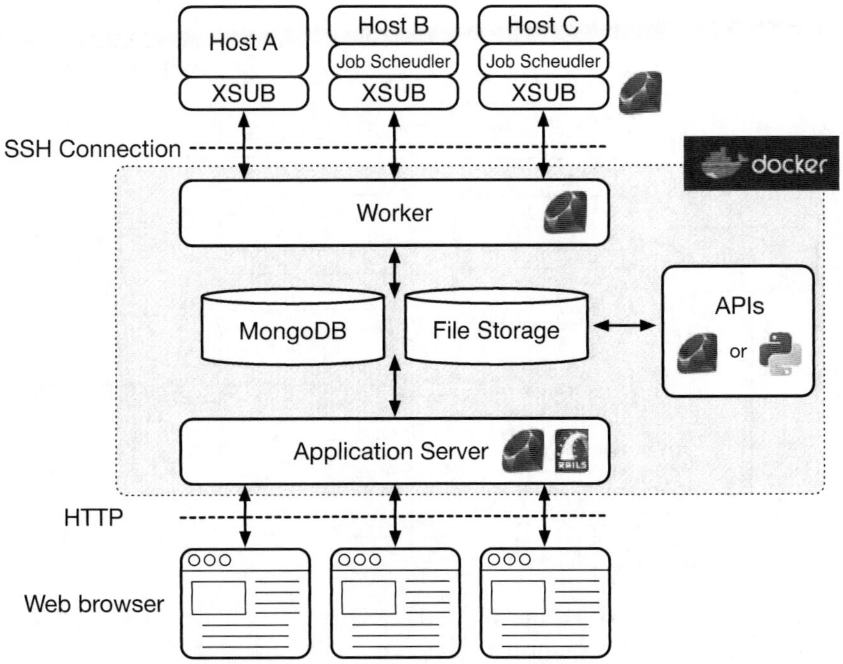


Figure 1: An overview of the system architecture of OACIS. The application server developed on Ruby on Rails framework provides a user interface. The request from users are stored in the MongoDB and they are handled by another daemon process called a "worker". The worker process makes an SSH connection to remote hosts and submits jobs. After the jobs finished, worker downloads results and stored them into the database and the file system. Users can also use Python or Ruby scripts to operate OACIS. The Ruby logo by Yukihiro Matsumoto is licensed under CC BY-SA 2.5.

Set", "Run") as shown in Fig. 2. Each simulator has several ParameterSets, and each ParameterSet has several Runs. A simulator corresponds to an executable command to run a simulation program. A ParameterSet represents a set of parameter values which are required by the simulator. Let us consider a simulation model which has two input parameters $p_1$ and $p_2$, for example. A set of values, say $\{p_1 = 0.2, p_2 = 0.5\}$, corresponds to a ParameterSet. A Run corresponds to a single Monte-Carlo run having a unique random number seed.

In addition to these three layered structures, you can define a post-process, called "Analyzer", which is conducted against simulation results, such as statistical analysis and visualization. For each Simulator, we can define several Analyzers. Likewise Simulators, Analyzers are executed at a computational host and their results are stored automatically after they finished. A result of Analyzer is called "Analysis". The relation between Analyzer and Analysis is similar to that between Simulator and Run. We can define two types of analyzers. One is analyzers conducted against the simulation result of a single Run. The other one is analyzers conducted against the simulation results for all the runs in a single ParameterSet. An Analysis are created under a Run or a ParameterSet depending on the type of the Analyzer.

### *2.2. Job Execution Sequence*

When you register a simulator on OACIS, you save a command line string to execute the simulation, not the execution program itself. By this specification, OACIS can run various

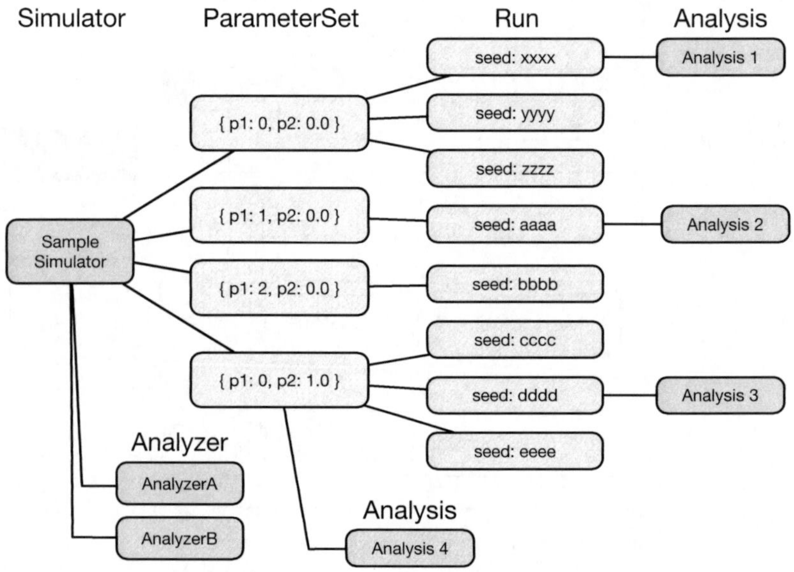


Figure 2: A hierarchical data structure is used by OACIS. On top layer, "Simulator" objects exists which corresponds to an executable program. "ParameterSet" corresponds to a set of parameter values. Under a ParameterSet, "Run" objects are stored, each of which has a unique random number seed. In addition to these, we can define a post process called "Analyzer".

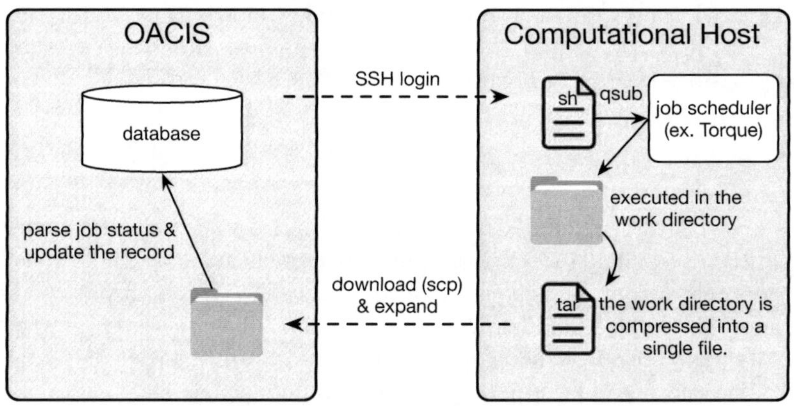


Figure 3: A sequence of job executions. OACIS logs in to a computational host via SSH. Then it makes a shell script and a temporary directory for the job. The shell script is submitted to a job scheduler and then the job is executed in the temporary directory. When the job finishes, the temporary directory is compressed into a single file. OACIS downloads the compressed file and expand it to store the simulation outputs. The logs of the job are also parsed by OACIS and recorded in the database.

programs written in any programming language. It also means that the simulation program must be compiled on computational hosts before submitting a job. When a job is submitted, OACIS generates a shell script, which we call "job script", including the command line to execute the program. For each Run, one job script is created. Job scripts created by OACIS are submitted to the job schedulers (such as Torque) on computational hosts via SSH. Just before submitting job scripts, a temporary directory called "work directory" is created for each job. A job is executed in its work directory and the files and directories created under this directory is included into OACIS as its simulation outputs. Thus, OACIS requires simulators to create all the output files under the current directory. The whole sequence is illustrated in Fig. 3.

When a job is executed, input parameters of the simulator are given as command line arguments or by a JSON file. We can choose one of these when registering the simulator on OACIS. If you choose the former one as a way to set input parameters, the parameters are given as the command line arguments, i.e., the command to execute the simulation program is concatenated by parameters and embedded into the job script. If you choose the JSON format as a way to set input parameters, a JSON file named *_input.json* is created in the work directory before execution of the jobs. Simulator must be implemented such that it reads the json file in the current directory.

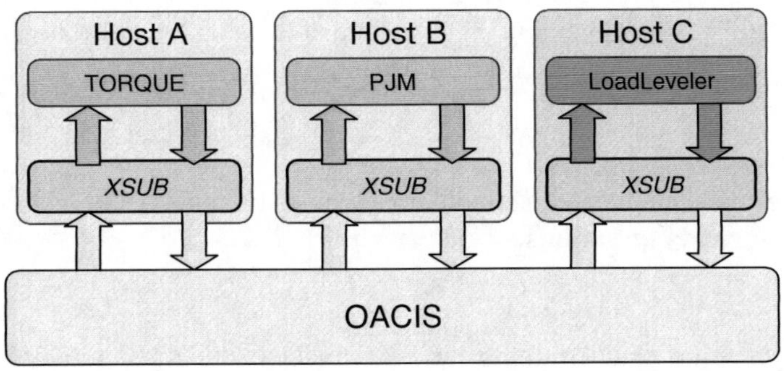


Figure 4: XSUB is a wrapper script for job schedulers. Even though the specifications of job schedulers are different for each host, OACIS has a uniform interface against the job schedulers using XSUB. XSUB is designed to be extensible so that it can adapt to various types of schedulers.

When using high-performance computers (HPCs), it is usually required to submit a job using a job scheduler such as Torque. However, the specifications of the job schedulers, not only the submission command but the format of job scripts, often vary according to systems. Job schedulers often have its own I/O format and we usually have to change job scripts to conform with its dialect. In order to absorb the difference of the schedulers, we introduced a small script called "XSUB" (`https://github.com/crest-cassia/xsub`). XSUB is a wrapper script to support various job schedulers and is also released as an open-source software under the MIT license. By introducing XSUB, OACIS does not need to prepare the script specialized for each job scheduler as shown in Fig. 4. A user can support new job scheduler without changing the code of OAICS, which makes the whole architecture more flexible and extensible.

### *2.3. Read-Only Mode*

When you want to share your results with the researcher at a remote site, "Read-Only" mode of OACIS is quite useful. If you launch OACIS in Read-Only mode, creating documents (Simulators, Parameter sets, Runs, Hosts, Analyzers, and Analyses) are prohibited, so it is not possible to submit a job or register a simulator. Therefore, you can run an OACIS server in a public network more safely.

A typical usage of Read-Only mode would be the following (See Fig. 5). A researcher executes simulations using OACIS launched at his/her local site. After the jobs finsihed, the researcher synchronizes the data to the remote OACIS which are launched at a cloud server in Read-Only

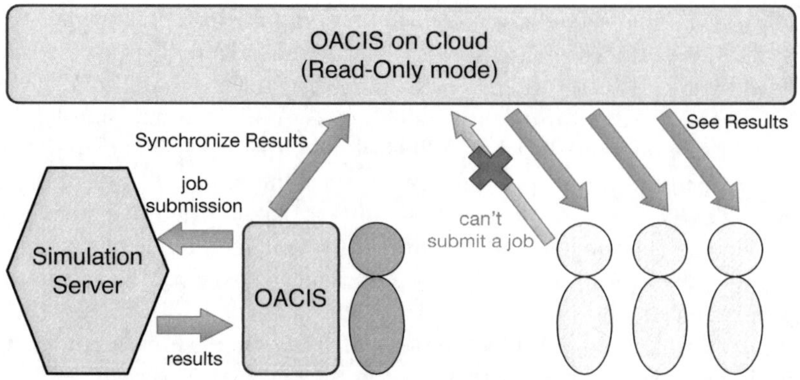


Figure 5: A use case of Read-Only mode. A researcher uses OACIS to manage jobs in the intranet. In order to share the results with the collaborators, the researcher launched another OACIS with Read-Only mode on a cloud server, and upload the results in the local OACIS. By sending a URL of the public OACIS, collaborators may access to it.

mode. The researcher may allow the collaborators to see the results while preventing a malicious user from executing invalid commands.

*2.4. Application Programming Interfaces*

OACIS provides application programming interfaces (APIs) in Ruby and Python programming languages. In principle, any operations on OACIS can be conducted by APIs, such as finding Simulators, creating a new ParameterSet, and deleting a Run. Actually, the same methods are used behind the web browser interfaces.

The following is a small sample conducting a parameter sweep over parameters *p1* and *p2* of a Simulator named *my_simulator*.

Listing 1: A sample script showing a usage of OACIS APIs.

```
# get Simulator and Host objects
sim = Simulator.find_by_name("my_simulator")
host = Host.find_by_name("localhost")

p1_values = [1.0,2.0,3.0,4.0,5.0]
p2_values = [2.0,4.0,6.0,8.0,10.0]

# iterate over p1 and p2
p1_values.each do |p1|
  p2_values.each do |p2|
    param = {"p1"=>p1,"p2"=>p2}
    # create a ParameterSet for the given parameters
    ps = sim.find_or_create_parameter_set( param )
    # create Runs under the given ParameterSet
    runs = ps.find_or_create_runs_upto(5, submitted_to: host)
  end
end
```

Using these APIs, we can automate explorations in parameter space. Applications include parameter sweeps, sensitivity analysis, optimization of parameters, and applying machine learning to the results of simulation results.

IOP Conf. Series: Journal of Physics: Conf. Series **921** (2017) 012001 doi:10.1088/1742-6596/921/1/012001

## 3. Getting Started with OACIS

### *3.1. Installation*

To start using OACIS, we need to set up both OACIS and computational hosts. There are two ways of setting up OACIS. The first one is running a virtual machine image in which OACIS is already set up using Docker (`https://www.docker.com/`). In the docker image, all the prerequisites are already installed. The image is built on "docker hub" (`https://hub.docker.com/`), a public site hosting docker images, which lets you start OACIS with a few commands. The second one is using a native (non virtual machine) environment and setup prerequisites manually. For the latter option, we need to use a Unix-like OS, such as Linux or macOS, and appropriate version of Ruby and MongoDB.

In this section, we briefly explain how to set up OACIS using Docker to let you try OACIS quickly. For the full instruction on the installation, please refer to the documentation of OACIS (http://crest-cassia.github.io/oacis/en/install.html).

First, install Docker on your machine. The installation procedure depends on your platform and the official documentation of Docker tells you how to install Docker. https://docs.docker.com/engine/installation/. After you installed Docker, run the following command to download the image and run the container. (If you are using Docker Toolbox, the command is different. Please refer to the documentation at the repository page (`https://github.com/crest-cassia/oacis_docker`).

```
docker run --name my_oacis -p 127.0.0.1:3000:3000 -dt oacis/oacis
```

When you run this command for the first time, it takes time to download the image from the public repository of docker images. After the image is downloaded, the container (a virtual machine) is launched and OACIS is started in the container. A few tens of seconds are necessary to complete the booting of OACIS. Access http://localhost:3000 to see the top page of OACIS as in Fig. 6.

Figure 6: A snapshot of the top page of OACIS.

### *3.2. Registering Hosts and Simulators*

After we installed OACIS, we need to set up computational hosts and register a simulator on OACIS.

To add a computational host, we first need to set up an SSH authentication key such that OACIS can connect to it without entering a password. Then XSUB must be installed to submit jobs. The detailed set up sequence are shown at a page of the OACIS documentation (`http://crest-cassia.github.io/oacis`). If you are using Docker, you can skip this process since the container is already set up also as a computational host.

To submit jobs using OACIS, we need to register a simulator on OACIS. Since we save the command line to execute the simulation, not the simulation program itself, simulators must be

IOP Conf. Series: Journal of Physics: Conf. Series **921** (2017) 012001 doi:10.1088/1742-6596/921/1/012001

prepared on the computational host in advance. In order to execute simulators from OACIS, your simulator must satisfy the following requirements.

(i) The output files or directories must be created in the current directory.
(ii) Simulator must receive input parameters either from command line arguments or from a JSON file.
(iii) The simulator must not conflict with the files *_status.json, _time.txt, _version.txt*, which are created by OACIS in the work directory.
(iv) The simulator must return 0 when finished successfully. The return code must be non-zero when an error occurs during the simulation.

The first requirement is about the simulation outputs. OACIS creates a temporary directory for each job and executes the job in that temporary directory. Since all the files and directories in the temporary directory are stored in OACIS as the simulation outputs, your simulators must create output files under the current directory. The second requirement specifies how simulators receives its input parameters given by OACIS. The third requirements is about the temporary files. To record the information of jobs, OACIS creates several files in the work directory. Simulators must not conflict with those. The fourth requirement is necessary to check if the simulator finished successfully or not. We prepared a sample simulator in (`https://github.com/yohm/sim_ns_model`).

After you prepared a simulator, you can register it on OACIS from a web browser interface. An instruction can be found at the tutorial page of OACIS documentation (`http://crest-cassia.github.io/oacis`). After simulators are registered, we can submit jobs. All the jobs are automatically managed by OACIS thus you can trace the simulation output even after several months. OACIS also provides a function to make a plot, with which we are able to visually inspect how the output depends on input parameters. (See Fig. 7 for examples.) The tutorial page gives you an instruction about a typical usage of OACIS using a simple sample simulator.

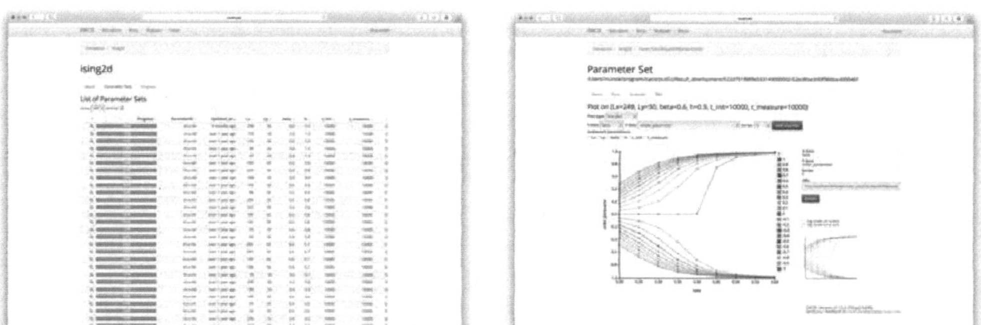


Figure 7: (left) A snapshot of the list of ParameterSets. (right) A snapshot of a plot made by OACIS. In this plot, a point corresponds to a ParameterSet. All the relevant ParameterSets are automatically collected by OACIS.

## 4. Conclusion

In this short paper, we presented a software framework OACIS. It is an ongoing project so the development will continue to enhance the functionality and improve the usability even more. If you are interested in this project, please check the latest status in our official project page (`https://github.com/crest-cassia/oacis`) and try to use it in your research.

**acknowledgments**
This research was supported by JST CREST and by MEXT as "Exploratory Challenges on Post-K computer(Studies of multi-level spatiotemporal simulation of socioeconomic phenomena)".

## References

[1] Murase Y, Uchitane T and Ito N 2014 *Physics Procedia* **57** 73–76
[2] Noda I, Ito N, Izumi K, Yamashita T, Mizuta H, Kamada T, Murase Y, Yoshihama S and Hattori H 2015 *2015 IEEE/WIC/ACM International Conference on Web Intelligence and Intelligent Agent Technology (WI-IAT)* vol 3 pp 22–25
[3] Murase Y, Török J, Jo H H, Kaski K and Kertész J 2014 *Phys. Rev. E* **90** 052810+
[4] Murase Y, Jo H H, Török J, Kertész J and Kaski K 2015 *PLoS ONE* **10** e0133005+
[5] Török J, Murase Y, Jo H H, Kertész J and Kaski K 2016 *Phys. Rev. E* **94** 052319
[6] Kuwabara A, Yoshioka N, Shimada T and Ito N 2016 *Journal of Physics: Conference Series* **750** 012018
[7] Torii T, Izumi K and Yamada K 2015 *Evol. Inst. Econ. Rev.* **12** 395–412

IOP Conf. Series: Journal of Physics: Conf. Series **921** (2017) 012002 doi:10.1088/1742-6596/921/1/012002

# Interactions between proteins and poly(ethylene-glycol) investigated using molecular dynamics simulations

**Giovanni Settanni, Jiajia Zhou, Friederike Schmid**
Institute of Physics, Johannes-Gutenberg University Mainz, Germany
E-mail: settanni@uni-mainz.de,friederike.schmid@uni-mainz.de

**Abstract.**
Poly(ethylene-glycol) (PEG) is a polymer used to coat therapeutic preparations, like drugs or drug nanocarriers, and improve their efficacy. This effect is probably due to a reduction of the interactions of the coated species with the host organism. Nevertheless, experiments show that PEGylated materials do interact with the surrounding biological milieu, and in particular with blood proteins. Here, we use atomistic molecular dynamics simulations to characterize the interactions between the polymer and several blood proteins. In these simulations, the proteins are immersed in a mixture of PEG and water molecules. We observe how PEG distributes around the protein surface and measure PEG-protein interactions in terms of preferential binding coefficients.

## 1. Introduction

Poly(ethylene-glycol) (PEG) is a non-toxic polymer highly soluble both in water and non-polar solvents and finds application in a broad variety of contexts, including for example, as dispersant in toothpastes, lubricant in skin creams and anti-foaming agent in the food industry. In biomedical applications, PEG, when attached to a therapeutic molecule or nanoparticle (PEGylation), confers them extra resistance to degradation, extended blood clearance time and, ultimately, improved therapeutic efficacy. These improvements are obtained by a combination of factors like increased overall volume, which prevents renal clearance, but also a reduction of unspecific interactions with the surrounding environment. This last phenomenon goes under the name of stealth effect, and results in the ability of PEG to screen the PEGylated species from the immune response of the host organism.

Although PEG helps reduce unspecific interactions with the surrounding biological milieu, it does not remove them completely. It has been shown that, also due the constant exposure to PEG-containing products, the organism can produce antibodies against PEG. In addition, many different proteins adsorb on PEG-coated nanoparticles, as soon as these enter in contact with blood[1]. A molecular understanding of these protein adsorption phenomena on PEG-coated surfaces and more in general of the interactions between PEG and proteins is still missing.

Atomistic molecular dynamics simulations can help to fill this gap. They have been successfully used to address protein adsorption on a large variety of surfaces [2, 3], as well as the behavior of specific blood proteins on model surfaces[4, 5]. While accurate atomistic models and force field already exist for many metallic, metal-oxides and some polymeric surfaces[3], the surface exposed by PEGylated species poses extra challenges due to its brushy nature. These challenges have been recently addressed by simulating the proteins of interest in a mixture of water and PEG molecules[6, 7]. With this approach, which is similar in spirit to what has been done in computational drug screening exercises[8, 9], the possible binding modes of PEG with the protein surface are sampled accurately and their relative importance is determined from the simulations.

Here, we extend the approach developed in [6] to additional blood proteins and use it to determine their preferential binding coefficient for PEG.

## 2. Methods

*Molecular dynamics simulations* All the simulations were carried out using the program NAMD [10] and the charmm27 force field[11] with the extension for PEG[12]. Tip3p[13] was used as model for the explicit treatment of water. An integration time step of 1 fs was used across the simulations. Simulations were carried out using periodic boundary conditions. A pressure of 1atm and a temperature of 300K were maintained constant during the simulations using the Langevin piston algorithm and Langevin thermostat [14, 15]. A cutoff of 1.2 nm was used for the non-bonded interactions with a switch function. Long range electrostatic interactions were treated using the smooth particle mesh Ewald (PME) method [16] with a grid spacing of about 0.1 nm. To prepare the PEG-water mixture, 64 PEG molecules ($H-[O-CH_2-CH_2]_n-OH$, with $n$ either 4 or 7) were placed on a 4x4x4 grid with 1.0 nm spacing between grid points. Then 1 ns high temperature (700K) simulations in vacuum with damped electrostatics interactions (dielectric constant 200) were run to randomize the initial dihedral distribution of the PEG molecules. The PEG molecules were then immersed in a box of water molecules and sodium and chlorine ions were added to reach physiological concentration (0.15 M). Mixtures with different concentrations of PEG were obtained by changing the size of the water box surrounding the PEG molecules. The prepared mixtures were then equilibrated first at high temperature (373K) for 1.0 ns and then at 300K for 1.0 ns. Several plasma proteins have been considered: human serum albumin(HSA), bovine serum albumin(BSA), transferrin(TF) and apolipoprotein A1 (ApoA1). The initial coordinates of the proteins were taken from the Protein Data Bank(PDB) (see Tab. 1 for the list of PDBids). Each protein was immersed in a box filled by replicating the coordinates of the PEG-water mixtures obtained before in the three space directions and removing mixture atoms in close contact with protein atoms. The final PEG concentration in the simulation boxes is reported in Tab. 1. The size of the boxes was large enough to leave at least 1.0 nm from each protein atom and the box boundary. In the case of HSA, larger box sizes with 1.5 and 2.0 nm distances between protein and box boundary were simulated to investigate the dependence of the simulation results upon box size. The total charge of the systems was further neutralized by changing an adequate number of water molecules into ions. The complete systems were then minimized using the steepest descent algorithm for 10000 steps with harmonic restraints on the heavy atoms of the proteins. Then the systems were equilibrated at room temperature and pressure for 1.0 ns during which the harmonic restraints were gradually removed and for 1.0 ns without restraints. Finally production runs were started with 4 or 5 replicas for each system. Length of the runs ranged from 100 to 200 ns (see tab. 1).

*Analysis of trajectories* The direct PEG-protein interactions along the simulations were measured using the NAMD pair-interaction utility. The time series along the trajectories of the number of PEG and water heavy atoms found within a distance $d$ of the protein or of each amino acid type was determined using the "pbwithin" selection command of VMD. The ratio of these numbers was compared to the ratio between all the PEG and water heavy atoms in the simulation box (bulk ratio). A PEG/water ratio larger than bulk for an amino acid implies the presence of an effective attractive interaction between PEG and the amino acid. On the other

IOP Conf. Series: Journal of Physics: Conf. Series **921** (2017) 012002 doi:10.1088/1742-6596/921/1/012002

**Table 1.** List of the simulations performed

| System | Box size (Å) | N. Atoms | PEG length | [PEG] (g/ml) | Simulation time (ns) |
|---|---|---|---|---|---|
| HSA+lipids | 103 | 114000 | 4 | 0.13 | 4 x 105 |
| HSA | 98.8 | 100881 | 4 | 0.08 | 4 x 100 |
| | 98.2 | 99301 | 4 | 0.11 | 4 x 100 |
| | 108.6 | 134134 | 4 | 0.12 | 4 x 125 |
| | 118.2 | 172541 | 4 | 0.12 | 5 x 100 |
| BSA | 103.6 | 116489 | 4 | 0.11 | 4 x 100 |
| Trasferrin | 108 | 132009 | 4 | 0.11 | 5 x 200 |
| | 108.7 | 134663 | 4 | 0.07 | 5 x 200 |
| | 108.4 | 133303 | 7 | 0.07 | 5 x 200 |
| | 108.8 | 134372 | 7 | 0.04 | 5 x 200 |
| ApoA1 | 192 | 739653 | 4 | 0.13 | 4 x $\approx$ 100 |

hand, amino acids with a PEG/Water ratio smaller than bulk exert an effective repulsion for PEG. For the residue specific PEG/Water ratio calculations we used $d = 5.0$Å. The preferential binding coefficient measures the excess of PEG found in the local environment of the protein, i.e., in its vicinity (Fig. 1) and was defined as:

$$\Gamma = [PEG]_L - [H_2O]_L \frac{[PEG]_B}{[H_2O]_B} \tag{1}$$

where $[\cdot]_L$ is the concentration of PEG or water in the vicinity of the protein and $[\cdot]_B$ the concentration in the bulk, that is, away from the protein. The local environment of the protein is defined as a shell of thickness $d$ around the protein. $\Gamma$ converges for $d \geq 9.0$Å, which is the value we used for these calculations.

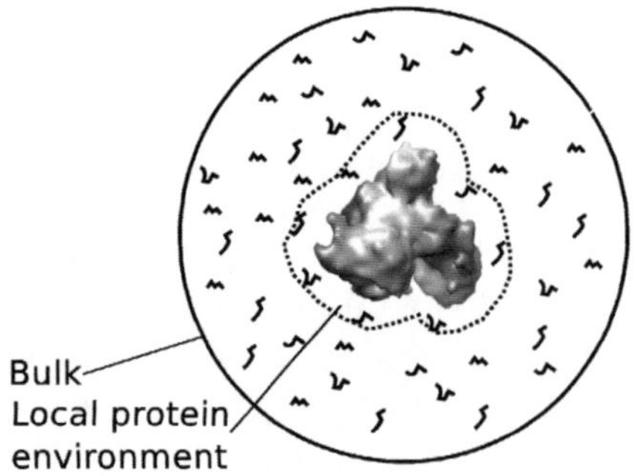


**Figure 1.** The concentration of PEG in the local environment of the protein and in the bulk solution are used for the measure of the preferential binding coefficient (Eq. 1).

## 3. Results and Discussion

Four different plasma proteins have been simulated immersed in a PEG/water mixture, as reported in section 2. Three of them (HSA, BSA, TF) are relatively abundant in plasma but not on the surface of PEGylated nanoparticles and one (ApoA1) is abundant on the nanoparticle [1]. The simulations were performed at different PEG concentrations to verify a possible dependence of the results on this parameter. Additional simulations explored the role of PEG length and box size. Each simulation was repeated 4 to 5 times to improve the statistical accuracy of the results. Only short PEG molecules were considered (4 or 7 monomers), which diffuse fast and allow for a thorough exploration of the possible binding modes with the protein. This insures

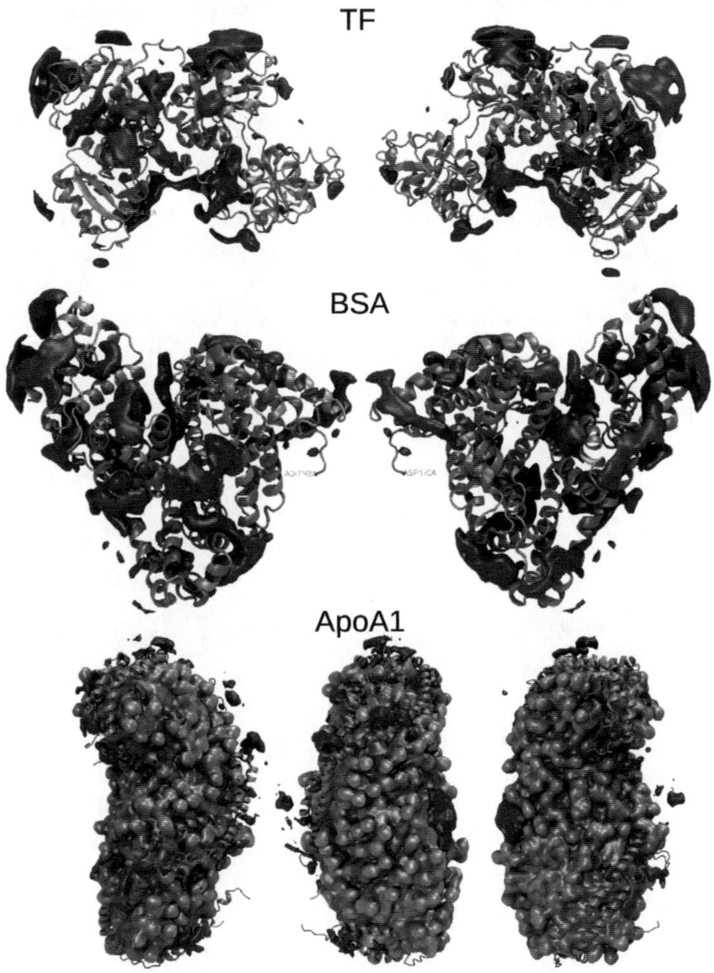


**Figure 2.** Cartoons of three different simulated proteins (cyan) where regions with PEG density larger than twice as in the bulk have been highlighted (dark blue). Left and right and center pictures represent different views of the proteins. Negatively charged residues are highlighted in red. In the case of ApoA1, the lipids present in the apolipoprotein are also rendered as a continuous surface.

that the observed quantities converge in the accessible simulation times (100ns-200ns). It is worth noting that, since the persistence length (3.8Å) of PEG is comparable with the distance between monomers, longer PEG chains are expected to explore the same binding modes as the shorter chains simulated here. The simulations show that the protein structures are not affected by the presence of PEG. The average root mean square deviation with respect to the starting structure or, in the case of ApoA1 to a control run in pure water, is in agreement with native state fluctuations of proteins of similar size (3-5Å for HSA, BSA and TF).

PEG molecules in the simulations reach and leave the surface of the protein continuously, in a rather dynamical equilibrium. However, some regions of the surface are more frequently populated by PEG (Fig. 2).

We, then, analyzed in more detail the possible origin of the observed uneven distribution of

PEG on the protein surface. We hypothesized that each amino-acid type may have a different affinity for PEG molecules. To test the hypothesis, we measured the average ratio between the number of PEG and water heavy atoms (PEG/Water ratio) in a 5.0Å shell surrounding each amino acid (Fig. 3). The data show that some amino acids are more likely to have a concentration of PEG higher than bulk, while others have a smaller concentration. In other words, some amino acids exert an effective attraction for PEG while other exert a repulsion. The PEG-amino acid interaction pattern look also very similar across the different simulations. Indeed, we verified that the the residue-specific PEG/Water ratio, normalized to the bulk value in each simulation, are highly correlated across the simulations, at least for the most exposed amino acids (Fig. 4). These information, obtained from similar simulations, allowed us to describe PEG-protein interactions with a simplified model based on the solvent accessibility of each amino acid on the protein surface[6].

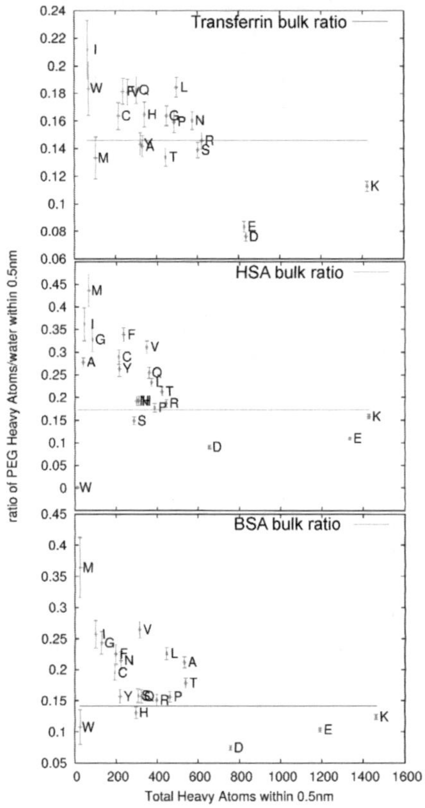


**Figure 3.** Ratio between PEG heavy atoms and water molecules within 5Å of each residue types vs sum of PEG+water (indicating the degree of exposure of each residues type). Negatively charged residues "repel" PEG.

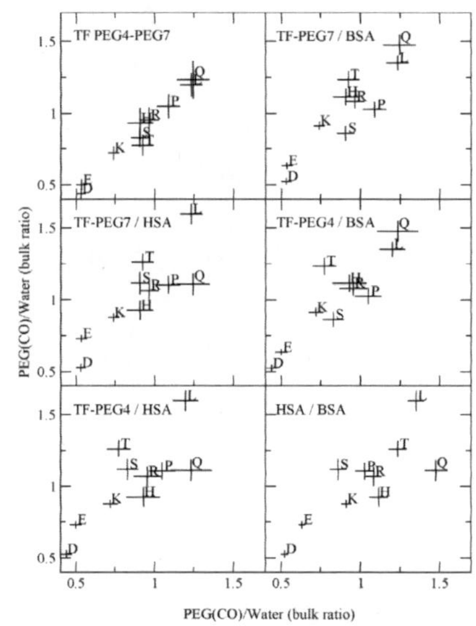


**Figure 4.** Correlation plots of the observed residue specific PEG/water ratios between the various simulations. Correlation coefficients all above 0.8 indicate the substantial independence of the values from the parameters of the simulations.

A further analysis of the data consists in the measurements of the preferential binding coefficient of each protein for PEG. Preferential binding coefficients are derived from the Kirkwood-Buff integrals of the pair correlation functions in a multi-component system[17]. In particular they measure the excess number of cosolvent molecules in the vicinity of each protein molecule, thus they are positive in case of a preferential binding of the cosolvent to the protein, and negative otherwise. Preferential binding coefficients are thermodynamic parameters that can be accessed experimentally through high precision densitometry experiments[18], differential scanning calorimetry[19] and vapor pressure osmometry[20]. In the present case, they report the affinity of each protein for the cosolvent (PEG) *versus* the solvent (water) and can be approximated using eq. 1[21]. The measurements we have performed using the simulation data (Tab. 2) show in most cases positive coefficients very close to zero, indicating a very weak preference of the analyzed proteins for PEG.

**Table 2.** Preferential binding coefficients measured along the simulations

| System | Box size (Å) | PEG length | [PEG] (g/ml) | Γ (n.mol.) |
|---|---|---|---|---|
| HSA+lipids | 103 | 4 | 0.13 | 3.4 (1.2) |
| HSA | 98.8 | 4 | 0.08 | 5.0 (0.8) |
| | 98.2 | 4 | 0.11 | 7.0 (1.3) |
| | 108.6 | 4 | 0.12 | 2.9 (0.9) |
| | 118.2 | 4 | 0.12 | -0.1 (1.1) |
| BSA | 103.6 | 4 | 0.11 | 0.0 (1.0) |
| Trasferrin | 108 | 4 | 0.11 | 4.8 (0.9) |
| | 108.7 | 4 | 0.07 | 7.5 (1.0) |
| | 108.4 | 7 | 0.07 | 4.7 (0.7) |
| | 108.8 | 7 | 0.04 | 2.9 (0.7) |
| ApoA1 | 192 | 4 | 0.13 | -1.2 (1.6) |

## 4. Conclusions

Here we show how atomistic molecular dynamics simulations can be used fruitfully to assess the interaction of plasma proteins with PEG. The simulations reveal that the interaction patterns with surface amino acids are similar across the different proteins and the different simulation conditions analyzed here. This fact has allowed for building simplified models of protein interactions with PEG[6]. The simulations provide also direct measurements of the preferential binding coefficients which could be assessed experimentally to test and help improve the available force fields. The approach used in the present work is rather general and could be extended to evaluate interactions of proteins with other polymers that, like PEG, are highly soluble in waters and do not induce large conformational changes to proteins. Indeed, large efforts are being made to identify polymers that could replace PEG in biomedical applications and solve the issues related to its possible long-term toxicity and nonbiodegradability. We expect the techniques developed here to help in that direction.

*Acknowledgments*
GS gratefully acknowledges financial support from the Max-Planck Graduate Center with the University of Mainz. We gratefully acknowledge support with computing time from the HPC facility Mogon at the university of Mainz and the High performance computing center Stuttgart. This work was supported by the German Science Foundation within SFB 1066 project Q1.

## References

[1] Schöttler S, Becker G, Winzen S, Steinbach T, Mohr K, Landfester K, Mailnder V and Wurm F R 2016 *Nat. Nanotechnol.* URL `http://dx.doi.org/10.1038/nnano.2015.330`

[2] Heinz H and Ramezani-Dakhel H 2016 *Chem. Soc. Rev.* **45** 412–448 ISSN 1460-4744 URL `http://pubs.rsc.org/en/content/articlelanding/2016/cs/c5cs00890e`
[3] Ozboyaci M, Kokh D B, Corni S and Wade R C 2016 *Q. Rev. Biophys.* **49** ISSN 0033-5835, 1469-8994 URL `https://doi.org/10.1017/S0033583515000256`
[4] Köhler S, Schmid F and Settanni G 2015 *PLOS Comput. Biol.* **11** 1–19 URL `http://dx.doi.org/10.1371/journal.pcbi.1004346`
[5] Köhler S, Schmid F and Settanni G 2015 *Langmuir* **31** 13180–13190 ISSN 0743-7463 URL `http://dx.doi.org/10.1021/acs.langmuir.5b03371`
[6] Settanni G, Zhou J, Suo T, Schttler S, Landfester K, Schmid F and Mailnder V 2017 *Nanoscale* **9** 2138–2144 ISSN 2040-3372 URL `http://pubs.rsc.org/en/content/articlelanding/2017/nr/c6nr07022a`
[7] Shao Q, He Y, White A D and Jiang S 2012 *J. Chem. Phys.* **136** 225101 URL `http://dx.doi.org/10.1063/1.4726135`
[8] Basse N, Kaar J L, Settanni G, Joerger A C, Rutherford T J and Fersht A R 2010 *Chem. Biol.* **17** 46–56 URL `http://dx.doi.org/10.1016/j.chembiol.2009.12.011`
[9] Seco J, Luque F J and Barril X 2009 *J. Med. Chem.* **52** 2363–2371 URL `http://dx.doi.org/10.1021/jm801385d`
[10] Phillips J C, Braun R, Wang W, Gumbart J, Villa E, Chipot C, Skeel R D, Kale L and Schulten K 2005 *J. Comput. Chem.* **26** 1781
[11] Mackerell A D, Feig M and Brooks C L 2004 *J. Comput. Chem.* **25** 1400–1415 ISSN 1096-987X URL `http://dx.doi.org/10.1002/jcc.20065`
[12] Lee H, Venable R M, Mackerell A D and Pastor R W 2008 *Biophys. J.* **95** 1590–1599 URL `http://dx.doi.org/10.1529/biophysj.108.133025`
[13] Jorgensen W L, Chandrasekhar J, Madura J D, Impey R W and Klein M L 1983 *J. Chem. Phys.* **79** 926–935 URL `http://link.aip.org/link/?JCP/79/926/1`
[14] Martyna G J, Tobias D J and Klein M L 1994 *J. Chem. Phys.* **101** 4177–4189 URL `http://link.aip.org/link/?JCP/101/4177/1`
[15] Feller S E, Zhang Y, Pastor R W and Brooks B R 1995 *J. Chem. Phys.* **103** 4613–4621 URL `http://link.aip.org/link/?JCP/103/4613/1`
[16] Essmann U, Perera L, Berkowitz M L, Darden T, Lee H and Pedersen L G 1995 *J. Chem. Phys.* **103** 8577–8593 URL `http://scitation.aip.org/content/aip/journal/jcp/103/19/10.1063/1.470117`
[17] Kirkwood J G and Buff F P 1951 *J. Chem. Phys.* **19** 774–777 URL `http://dx.doi.org/10.1063/1.1748352`
[18] Lee J C and Timasheff S N 1981 *J. Biol. Chem.* **256** 7193–7201 ISSN 0021-9258, 1083-351X URL `http://www.jbc.org/content/256/14/7193`
[19] Poklar N, Petrovi N, Oblak M and Vesnaver G 1999 *Protein Sci.* **8** 832–840 ISSN 1469-896X URL `http://onlinelibrary.wiley.com/doi/10.1110/ps.8.4.832/abstract`
[20] Courtenay E S, Capp M W, Anderson C F and Record M T 2000 *Biochemistry* **39** 4455–4471 ISSN 0006-2960 URL `http://dx.doi.org/10.1021/bi9928871`
[21] Baynes B M and Trout B L 2003 *J. Phys. Chem. B* **107** 14058–14067 ISSN 1520-6106 URL `http://dx.doi.org/10.1021/jp0363996`

IOP Conf. Series: Journal of Physics: Conf. Series **921** (2017) 012003 doi:10.1088/1742-6596/921/1/012003

# Universality and phase diagrams of the Baxter-Wu Model in a Crystal Field: spin-1 and spin-3/2

**D. A. Dias**[1,2], **J. C. Xavier**[3] **and J. A. Plascak**[1,4,5]

[1]Departamento de Física, Instituto de Ciências Exatas, Universidade Federal de Minas Gerais, C.P. 702, 30123-970 Belo Horizonte, MG - Brazil
[2] Universidade Federal de Uberlândia, Campus Patos de Minas, Av. Getúlio Vargas 230, 38700-103 Patos de Minas, MG - Brazil
[3] Instituto de Física, Universidade Federal de Uberlândia, C.P. 593, 38400-902 Uberlândia, MG - Brazil
[4] Universidade Federal da Paraíba, Centro de Ciências Exatas e da Natureza - Campus I, Departamento de Física - CCEN Cidade Universitária 58051-970 - João Pessoa, PB - Brazil
[5]Center for Simulational Physics, University of Georgia, 30602 Athens, GA - USA

E-mail: dani_adias@yahoo.com.br,jcxavier@ufu.br,pla@hal.physast.uga.edu

**Abstract.** Conventional finite-size scaling and conformal invariance theory are used in order to study the critical behavior of the spin-1 and spin-3/2 Baxter-Wu model. For spin-1 the results are similar to the Blume-Capel model. However, for spin-3/2, the phase diagram is much richer, and presents, besides a pentacritical point, an additional multicritical endpoint. In both cases, the universality class is the same as the spin-1/2 model, even at the multicritical points.

The Baxter-Wu model in a crystal field can be defined by following Hamiltonian

$$H = -J \sum_{<ijk>} \sigma_i \sigma_j \sigma_k + D \sum_{i=1}^{N} \sigma_i^2, \tag{1}$$

where $J$ is the exchange interaction, the first sum extends over all elementary triangles of a triangular lattice with $N$ sites, the classical spin variables $\sigma_i$ take values $\sigma_i = -S, -S+1, ..., S-1, S$, with $S = 1/2, 1, 3/2, ...$, and $D$ is the crystal field or single ion anisotropy. The above model for $S = 1/2$ [1] has been exactly solved [2] and the exponents are in the 4-states Potts model universality class [3, 4]. However, for $S \geq 1$ no exact results are available and even the phase diagram topology is not yet well resolved.

In order to study the critical behavior of the model given by (1) we have employed conventional finite-size scaling and conformal invariance theory through the transfer matrix formalism for infinite strips with finite width $L$ and periodic boundary conditions in all directions (for more details see [5, 6]). In summary, the second-order transition lines are obtained from

$$G_L(K,d)L = G_{L+3}(K,d)(L+3), \quad L = 3, 6, ..., \tag{2}$$

where $G_L(K,d)$ is the inverse of the correlation length and is given by

$$G_L(K,d) = \ln\left(\Re(\Lambda_L^1)/\Re(\Lambda_L^2)\right).$$

In the above equations we define $K = \beta J$, $d = D/J$, $\beta = 1/k_BT$, with $k_B$ being the Boltzmann constant, and $\Re[\Lambda_L^i(K,d)]$ is the real part of the $i$-th largest eigenvalue of the corresponding (non-Hermitian) transfer matrix. Multicritical points are obtained using a heuristic method in such a way that

$$G_L(K,d)L = G_{L+3}(K,d)(L+3) = G_{L+6}(K,d)(L+6), \quad L = 3, 6, ... \tag{3}$$

Note that Eqs. (2) and (3) are restricted to multiples of 3 to preserve the invariance of the Hamiltonian (1) under the reversal of all spins on any two of the three sub-lattices (each vertex of the triangle of the lattice can be viewed of belonging to three different sublattices in such a way that inverting the spins on any two of these sublattices leaves the Hamiltonian invariant). First-order transition lines are estimated by looking at the minimum of the following gap function

$$\Delta_p(K,d) = \Re[\Lambda_L^1(K,d)] - \Re[\Lambda_L^p(K,d)] \ , \tag{4}$$

where $p$ is the number of coexisting phases along the first-order transition. The conformal anomaly $c$ can be calculated from the large-$L$ behavior of the ground-state eigenvalue

$$\frac{\ln \Re(\Lambda_L^1)}{L} = \epsilon_\infty + \frac{\pi c v_s}{6L^2} + o(L^{-2}), \tag{5}$$

where $\epsilon_\infty = \frac{\ln \Re(\Lambda_L^1)}{L}$ is the bulk limit when $L \to \infty$ and $v_s = \sqrt{3}/2$ is the sound velocity. On the other hand, the scaling dimensions $x(n)$ can be obtained by extrapolating the sequence

$$x_L(n) = \frac{L}{\pi\sqrt{3}} \ln\left(\frac{\Re(\Lambda_L^1)}{\Re(\Lambda_L^n)}\right). \tag{6}$$

The corresponding critical exponents of the correlation function and correlation length are, respectively, given by $x(2) = \eta/2$ and $x(3) = 2 - 1/\nu$. The numerical diagonalization of the transfer matrix has been done by using the Lanczos method for non-Hermitian matrices [7] and, when needed, full diagonalization algorithms.

## Spin $S = 1$

For the spin-1 model, within our computational power, we can consider strip widths up to $L = 12$, and the results are shown in Figs. 1. It is interesting to look at Fig. 1 (a) where we have the whole set of the results obtained by considering all the strips for this model, which includes second-order transitions, first-order transitions, pentacritical points, as well as the thermodynamic limit extrapolation. Concerning the thermodynamic limit, we should emphasize that we have taken into account corrections to finite-size scaling. One can note that the first-order transition lines tend to meet the second-order transition lines at the pentacritical point. From the above results there is a strong indication that the spin-1 Baxter-Wu model shows a second-order transition line separated from a first-order transition line by a pentacritical point. Although this scenario is in disagreement with a picture from a previous paper by Kinzel et al. [8] (where the second-order transition line appears only in the limiting case $d \to -\infty$), the phase diagram has the same topology of the spin-1 Blume-Capel model [9, 10]. In the present case, the best estimate of the pentacritical point for the spin $S = 1$ model is $t_p^1 = 1.4020$ and $d_p^1 = 0.890254$.

The central charge has also been computed and the results support that the critical behavior of the critical line is described by a conformal field theory with $c = 1$. In addition, our results for the critical exponents are consistent with $\nu = 2/3$ and $\eta = 1/4$, which are the same of the pure Baxter-Wu model, even at the pentacritical point, meaning that the model belongs to the same universality class as the spin $S = 1/2$ model.

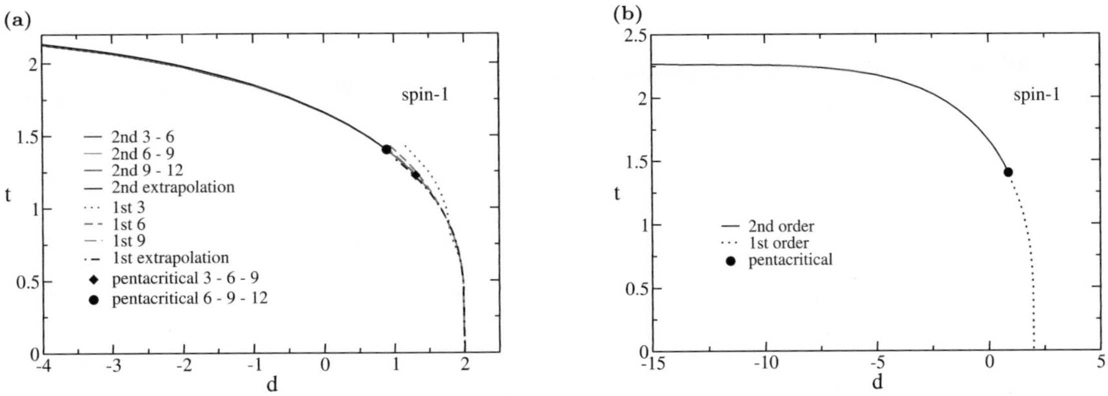


Figure 1: ( a) first-(broken lines) and second-order (continuous lines) transition temperatures as a function of the crystalline field $d$ for the spin-1 Baxter-Wu model. The solid dots are the pentacritical points. The numbers indicate the corresponding lattice sizes. Note that in the scale of the figure the curves are almost collapsed (except for the first-order transition lines close to the pentacritical point). ( b) spin-1 global phase diagram in the thermodynamic limit.

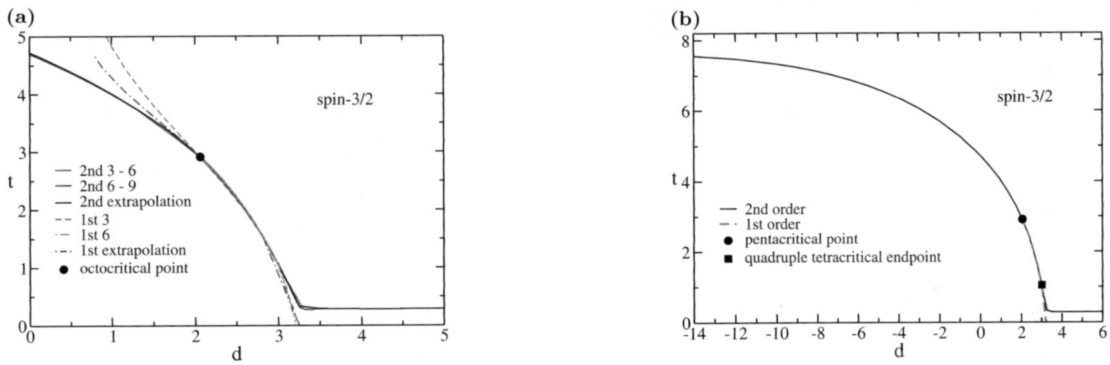


Figure 2: The same as Fig. 1 for spin-3/2.

## Spin $S = 3/2$

Regarding the spin-3/2 model we are limited to the width size $L = 9$, but we expect to obtain good results by taking $L = 3, 6$ and 9. The results are shown in Figs. 2. As before, in Fig. 2 (a) we have the whole set of results obtained by considering all the strips. At a first look, it seems that we have an octocritical point. However, by closely analysing the data, we can also expect to have another multicritical point at lower temperatures (although we cannot rule out completely the presence of an octocritical point). This extra multicritical point turns out to be a quadruple tetracritical endpoint, and is located at $t_{tce} = 1.05(5)$ and $d_{tce} = 3.01(5)$, while the previous one is a pentacritical point and is located at $t_p^{3/2} = 2.0620$ and $d_p^{3/2} = 2.9145$. The global phase diagram, in the thermodynamic limit, is shown in Fig. 2 (b). Independent of each above scenarios, one can say that the spin $S = 3/2$ Baxter-Wu model has a different phase diagram as the Blume-Capel, where the latter one presents only a double critical endpoint below the second-order transition line [11, 12, 13, 14, 15].

We have also computed the critical exponents and the central charge. The latter quantity is shown in Fig. 3. One can clearly see that the central charge, in the thermodynamic limit, is

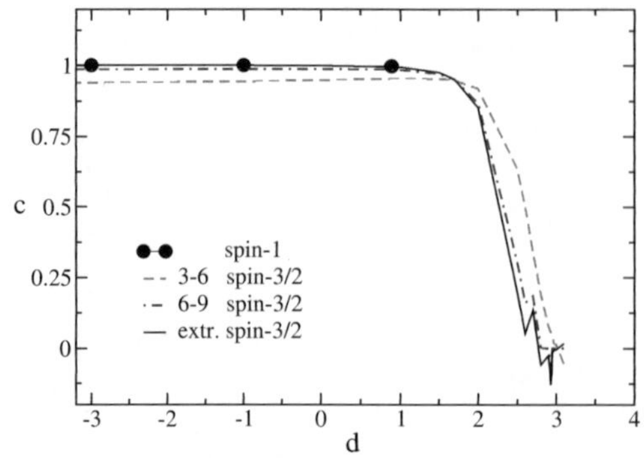


Figure 3: Central charge as a function of the crystal field. Full dots are some extrapolated results for spin-1 up to the pentacritical point, and the lines correspond to the results for spin-3/2.

also close to 1 (and comparable to those from the spin $S = 1$ model), but drastically changes in the region where a possible first-order transition may occur. In fact, for first-order transitions the central charge $c$ is zero (note that $c$ indeed oscillates around zero close to the multicritical endpoint). The critical exponents $\nu$ and $\eta$ are also compatible with those from the pure model.

## Acknowledgments

We would like to acknowledge Prof. J. Yeomans, Prof. J. L. Cardy and Prof. S. L. de Queiroz for fruitful discussions, and Prof. S.-H. Tsai and Prof. G. Weber for the computational help. This work was supported by CNPq, CAPES and FAPEMIG, Brazilian agencies, and also a CNPq grant 402091/2012-4.

## References

[1] D. W. Wood and H. P. Griffiths, *J. Phys. C* **5**, L253 (1972).
[2] R. J. Baxter and F. Y. Wu, *Phys. Rev. Lett.* **31**, 1294 (1973); R. J. Baxter and F. Y. Wu, *Aust. J. Phys.* **27**, 357 (1974); R. J. Baxter, *Aust. J. Phys.* **27**, 369 (1974).
[3] F. C. Alcaraz and J. C. Xavier, *J. Phys. A: Math. Gen.* **30**, L203 (1997).
[4] F. C. Alcaraz and J. C. Xavier, *J. Phys. A: Math. Gen.*, **32**, 2041-2060 (1999).
[5] M. L. M. Costa, J. C. Xavier, and J. A. Plascak, Phys. Rev. B **69**, 104103 (2004).
[6] D. A. Dias, J. C. Xavier, and J. A. Plascak, Phys. Rev. E **95**, 012103 (2017).
[7] G. H. Golub and C. F. van Loan, *Matrix Computations*, 3rd ed. (The Johns Hopkins University Press, Baltimore, 1996).
[8] W. Kinzel, E. Domany and A. Aharony, *J. Phys. A: Math. Gen.* **14**, L417 (1981).
[9] M. Blume, *Phys. Rev.* **141**, 517 (1966); H. W. Capel, *Physica* **32**, 966 (1966).
[10] J. A. Plascak, J. G. Moreira and F. C. Sá Barreto, *Phys. Lett. A* **173**, 360 (1993).
[11] J. A. Plascak and D. P. Landau, Computer Simulation Studies in Condensed Matter Physics XIII, D. P. Landau, S. P. Lewis and H.-B. Shuttler (eds) (Springer, Berlin, Heidelberg 2000).
[12] J. A. Plascak and D. P. Landau, Phys. Rev. E **67**, R015103 (2003).
[13] E. Costabile, J. Roberto Viana, J. Ricardo de Sousa and J. A. Plascak, *Physica A* **393**, 297 (2014).
[14] M. El Bouziani and A. Gaye, *Physica A* **392** 2643 (2013).
[15] J. C. Xavier, F. C. Alcaraz, D. P. Lara , and J. A. Plascak, *Phys. Rev. B* **57**, 11575 (1998).

IOP Conf. Series: Journal of Physics: Conf. Series **921** (2017) 012004 doi:10.1088/1742-6596/921/1/012004

# The zeros of the Energy Probability Distribution - A new way to study phase transitions -

**B. V. Costa**
Laboratório de Simulacão, Departamento de Física, ICEx
Universidade Federal de Minas Gerais, 31720-901 Belo Horizonte, Minas Gerais, Brazil
E-mail: bvc@fisica.ufmg.br

**L. A. S. Mól**
Laboratório de Simulacão, Departamento de Física, ICEx
Universidade Federal de Minas Gerais, 31720-901 Belo Horizonte, Minas Gerais, Brazil
E-mail: lucasmol@fisica.ufmg.br

**J. C. S. Rocha**
Departamento de Física, ICEB, Universidade Federal de Ouro Preto, 35400-000 Ouro Preto, Minas Gerais, Brazil
E-mail: jcsrocha@iceb.ufop.br

**Abstract.** We describe an iterative method to study the critical behavior of a system based on the partial knowledge of the complex Fisher zeros set of the partition function. The method is general with advantages over most conventional techniques since it does not need to identify any order parameter a priori. The critical temperature and exponents can be obtained with great precision. To test the method and to show how it works we applied it to the $2D$ Ising and $6-states$ Potts models. The strategy can easily be adapted to any model, classical or quantum, once we can build the corresponding energy probability distribution.

## 1. Introduction

In this communication, we present a new, general and efficient method based on the partial knowledge of the energy probability distribution ($EPD$) conjugated with an analogue of the Fisher zeros expansion of the partition function that allow us to precisely determine the transition temperature and exponents for a given system. In order to present the method we first introduce shortly the Fisher's zeros approach. Next we describe how it can be combined with the $EPD$ to obtain the critical behavior of a given system. To demonstrate the robustness of the method and its reliability our calculations are compared with the exact results for the $2D$ Ising model and the exactly known critical temperature for the $2D$ $6-states$ Potts model.

## 2. Fisher Zeros

Fisher has shown how the partition function can be written as a polynomial in terms of the variable $x = e^{-\beta\epsilon}$, where $\beta = 1/k_BT$ is the inverse of the temperature, $T$, $k_B$ is the Boltzmann constat, and $\epsilon$ is the energy difference between two consecutive energy states [1, 2, 3]. For a finite system, all roots of the polynomial lie in the complex plane. The coefficients of the polynomial are real implying that their roots appear in conjugate pairs. If the system under consideration undergoes a phase transition at a temperature $T_c$, the corresponding zero, $x_c$, must be real in the thermodynamic limit. To make those statements clearer we recall that the partition function is given by

$$Z = \sum_{\mathbf{E}} g\left(\mathbf{E}\right) e^{-\beta \mathbf{E}} = e^{-\beta\varepsilon_0} \sum_{n=1}^{N} g_n e^{-\beta n\varepsilon}, \tag{1}$$

where it is assumed that the energy, $\mathbf{E}$, can be written as a discrete set $\{\mathbf{E}_n = n\varepsilon\}; n = 0, 1, 2, \cdots$ and $\varepsilon_0$ is some constant energy threshold.

As pointed above, if the system undergoes a phase transition at $T_c$ the corresponding zero $x_t(L)$ moves toward the positive real axis as the system size grows. From now on this zero will be named dominant or the leading zero. In Fig. 1 it is shown a partial view of the distribution of zeros for the $2D$ Ising model in a square lattice of size $L = 32$. The dominant zero is identified in the figure. We observe that surprisingly the zeros do not appear in conjugated pairs as expected.

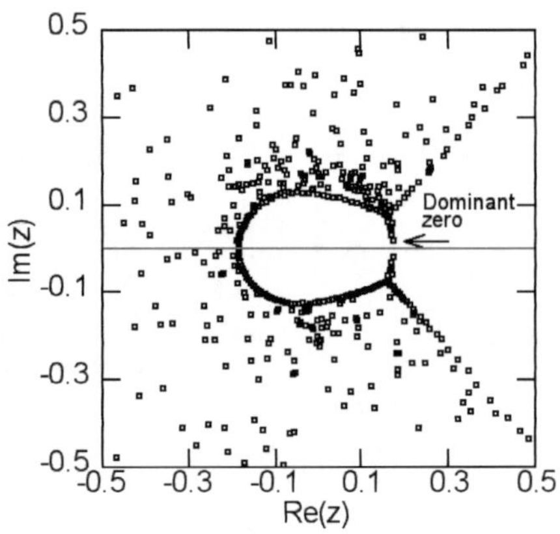


**Figure 1.** Partial view of the Fisher zeros distribution for the two dimensional Ising model for a $32 \times 32$ square lattice. We used the Mathematica's R solve package to get the zeros. Surprisingly, the expected conjugate pairs of zeros were not found. A discussion is found in the text. The arrow marks the dominant zero.

A closer analysis of $Z$ shows that the polynomial coefficients span over many orders of magnitude. For instance, for a lattice size $96 \times 96$ in the $2D$ Ising model one has 9217 possible energy values with the smallest and largest polynomial coefficient being $g(-18432) = 2$ and $g(0) = 2.304 \times 10^{2772}$ respectively. The large ratio between the coefficients turns the zeros finding into a formidable task even for known good algorithms. It should be interesting if we could focus our attention only onto the region where the dominant zero is located. This, in fact, can be done in a systematic way.

## 3. Energy Probability Distribution Zeros

If we multiply Eq. 1 by $1 = e^{-\beta_0 \mathbf{E}} e^{\beta_0 \mathbf{E}}$ it is rewritten as

$$Z_{\beta_0} = \sum_{\mathbf{E}} h_{\beta_0}\left(\mathbf{E}\right) e^{-\mathbf{E}\Delta\beta}, \tag{2}$$

where $h_{\beta_0}\left(\mathbf{E}\right) = g\left(\mathbf{E}\right) e^{-\beta_0 E}$ and $\Delta\beta = \beta - \beta_0$. Defining the variable $x = e^{-\varepsilon\Delta\beta}$ we obtain

$$Z_{\beta_0} = e^{-\varepsilon_0 \Delta\beta} \sum_{n} h_{\beta_0}(n) x^n, \tag{3}$$

where $h_{\beta_0}(n) = h_{\beta_0}(\mathbf{E}_n)$ is nothing but the non-normalized canonical energy probability distribution ($EPD$), hereafter referred to as the energy histogram at temperature $\beta_0$. There is a one to one correspondence between the Fisher zeros and the $EPD$ zeros since they are related through a trivial conformal transformation. Constructing the histogram at the transition temperature, i.e., $\beta_0 = \beta_c$, the dominant zero will be at $x_c = 1$, i.e., $Z = 0$ at the critical temperature ($\Delta\beta = 0$) in the thermodynamic limit. For finite systems, however, a small imaginary part of $x_c$ is expected. Indeed, we may expect that the dominant zero is the one with the smallest imaginary part on the real positive region regardless $\beta_0$. Once we locate the dominant zero its distance to the point $(1, 0)$ gives $\Delta\beta$ and an estimate for $\beta_c$.

For temperatures close enough to $\beta_c$ only states with non-vanishing probability to occur are pertinent to the phase transition. Thus, for $\beta_0 \approx \beta_c$ we can judiciously discard *small* values of $h_{\beta_0}$. Therefore, the zeros finder does not have to deal with high degree polynomials with coefficients spanning over prohibitive many orders of magnitude. Moreover, one may speculate that the dominant zero acts as an accumulation point such that even far from $\beta_c$ fair estimates can be obtained.

With this in mind we can develop a clear criterion to filter the important region in the energy space were the most relevant zeros are located. The idea follows closely the well known *Regula Falsi* method for solving an equation in one unknown. The reasoning is as follows: We first build a normalized histogram $h_{\beta_0^0}$ ($\text{Max}(h_{\beta_0^0}) = 1$) at an initial (False) guess $\beta_0^0$. Afterward, we construct the polynomial, Eq. 3, finding the corresponding zeros. By selecting the dominant zero, $x_c^0$, we can estimate the pseudo critical temperature, $\beta_c^0$. Regarding that $\beta_c(L)$ is the true pseudo-critical temperature for the system of size $L$, if the initial guess $\beta_0^0$ is far from $\beta_c(L)$ the estimative $\beta_c^0$ will not be satisfactory. Nevertheless, we can proceed iteratively making $\beta_0^1 = \beta_c^0$, building a new histogram at this temperature and starting over. After a *reasonable* number of iterations we may expect that $\beta_c^j$ converges to the true $\beta_c(L)$ and thus $x_c^j$ approaches the point $(1, 0)$. This corresponds to apply a sequence of transformations, $\mathfrak{T}$, such that $\beta^{n+1} = \mathfrak{T}\beta^n$. The transition temperature corresponds to the fixed point $\beta_c = \mathfrak{T}\beta_c$. The property $x_c^j \to (1, 0)$ can be used as a consistency check in this iterative process. An algorithm following those ideas is:

(i) Build a single histogram $h_{\beta_0^j}$ at $\beta_0^j$.

(ii) Find the zeros of the polynomial with coefficients given by $h_{\beta_0^j}$.

(iii) Find the dominant zero, $x_c^j$.
 a) If $x_c^j$ is close enough to the point $(1, 0)$, stop.
 b) Else, make $\beta_0^{j+1} = -\frac{\ln(\Re e\{x_c^j\})}{\varepsilon} + \beta_0^j$ and go back to 1.

In all our numerical results we observed that the choice of the starting temperature as irrelevant. To build the single histogram we follow the recipe given by Ferrenberg and Swendsen [4, 5]. To specify the energy subspace, the procedure is straightforward dealing with continuous transitions. In a first order transition the probability distribution is bi-modal. The energy distribution between the two peaks may reach low values, even lower than our cutoff criteria. In this case the procedure to define the relevant energy interval is subtler. A general way for dealing with this is to cut the low and high energy tails of the probability density preserving the important intermediate region.

## 4. Testing the algotrithm

### *4.1. 2D Ising Model*

The convergency and accuracy of the method can be tested by using the well-known Ising model

**Table 1.** Estimate of the pseudo critical temperature for the ferromagnetic 2$d$ Ising Model. The last line is the exact value up to five figures.

| L | $\beta$ | $T$ | $\|x\|$ | $\Im m(x)$ |
|---|---|---|---|---|
| 4 | 0.6667 | 3.0000 | 0.6262 | 0.7652 |
| | 0.4504 | 2.2202 | 1.1130 | 1.2180 |
| | 0.4236 | 2.3607 | 0.9752 | 1.0939 |
| | 0.4299 | 2.3261 | 1.0059 | 1.1218 |
| | 0.4284 | 2.3343 | 0.9985 | 1.1150 |
| | 0.4288 | 2.3321 | 1.00045 | 1.1168 |
| | 0.4287 | 2.3326 | 0.9999 | 1.1159 |
| 8 | 0.4287 | 2.3326 | 0.9924 | 0.4241 |
| | 0.4306 | 2.3223 | 0.9999 | 0.4273 |
| 16 | 0.4306 | 2.3223 | 0.9923 | 0.2088 |
| | 0.4325 | 2.3121 | 0.9999 | 0.2104 |
| 32 | 0.4325 | 2.3121 | 0.9872 | 0.1035 |
| | 0.4357 | 2.2952 | 0.9998 | 0.1049 |
| | 0.43575 | 2.2949 | 0.9999 | 0.1049 |
| 64 | 0.43575 | 2.2949 | 0.9911 | 0.0520 |
| | 0.4380 | 2.2831 | 1.00005 | 0.0525 |
| 96 | 0.43575 | 2.2949 | 0.9967 | 0.0349 |
| | 0.4388 | 2.2789 | 0.9999 | 0.0350 |
| estimate | 0.4404 | 2.2707 | - | - |
| $\infty$ | 0.4407 | 2.2691 | - | - |

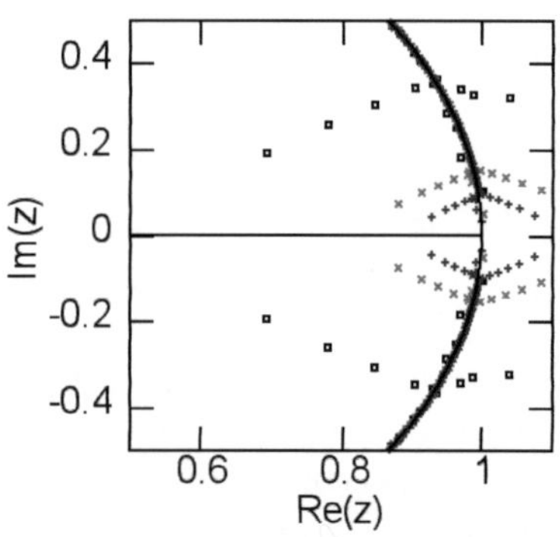


**Figure 2.** Plot section of the zeros distribution for the Ising model. Lattices sizes are $L = 32, 64, 96$ ($\square, \times$ and $+$ respectively.). The solid black line marks the perimeter of the unit circle and the real axis.

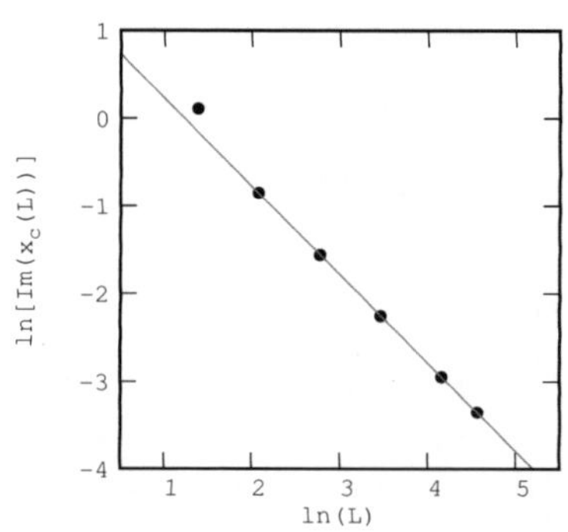


**Figure 3.** Log-log plot of $\Im m(x_c(L))$ versus $L$ for the Ising model. According to finite size scaling theory we expect the slope to be $\frac{1}{\nu} = 1$. From our data we obtained $\frac{1}{\nu} = 1.006(3)$.

[6] in a square lattice defined by the Hamiltonian

$$H = -J \sum_{<i,j>} \sigma_i \sigma_j, \tag{4}$$

where $J$ is a ferromagnetic coupling constant, $< i, j >$ means that the summation runs over nearest neighbors on a regular lattice and $\sigma_i = \pm 1$ is the spin variable at site $i$. The model has a continuous phase transition at $T_c = \frac{2}{\ln(1+\sqrt{2})} \frac{J}{k_B} \approx 2.269 \frac{J}{k_B}$ and was exactly solved in 1944 [6], being thus one of the preferred model for testing new methods on phase transitions. In Tab. 1 we show the results of the application of our method to the Ising model for several lattice sizes ($4 \leq L \leq 96$). The results were obtained by using the exact density of states calculated using the Mathematica® software with the code developed in Ref. [7].The interval of energy used was $[E_{min}, E_{max}]$ chosen in such way that for the normalized $h_\beta(n)$, both $h_\beta(E_{min})$ and $h_\beta(E_{max})$ are of order of $10^{-2}$. The zeros of the polynomial were calculated using the Mathematica® function *Solve*. In the last two lines of Tab. 1 we show the critical temperature result obtained from finite size scaling and the exact value up to 5 figures. The error is smaller than 0.5%. The zeros distribution is shown in Fig. 2. The critical exponent $\nu$ can straightaway be obtained. Finite size scaling theory predicts that $T_c(L) \sim T_c + aL^{-1/\nu}$ [8] and thus, we might expect that $x_c(L) \sim x_c + bL^{-1/\nu}$. Once $\Re e\{x_c(L)\} \approx 1, \quad \forall L$, the imaginary part of $x_c(L)$ should also scale as $L^{-1/\nu}$. In Fig. 3 a log-log plot of $\Im m(x_c(L))$ versus $L$ is shown for $L = 4, 8, 16, 32, 64, 96$. Discarding the smaller lattice size we obtain $1/\nu = 1.006(3)$, very close to the exact value, 1. Of course, better results for $T_c$ and $\nu$ are obtained by diminishing the energy threshold and using larger lattice sizes.

*4.2. 2D 6 − states Potts model*

The $q$ states Potts Model [9] is defined by the Hamiltonian,

$$H = -J \sum_{<i,j>} \delta_{\sigma_i, \sigma_j}, \tag{5}$$

where $J$ is a ferromagnetic coupling constant, $< i, j >$ stands for nearest neighbors, $\delta_{\alpha,\beta}$ is the Kronecker delta ($\delta_{\alpha\beta} = 1$ if $\alpha = \beta$, 0 otherwise) and $\sigma_i = 1, 2, \cdots, q$ is a spin like variable. The $q = 6$ states Potts model is known to have a first order transition at $T_c = \frac{1}{\ln(1+\sqrt{q})} = 0.80761....$ As in the Ising case our method applies straightforwardly. Table 2 shows in detail our results. The finite size scaling in this case has shown to be more subtle. For this weak first order

**Table 2.** Monte Carlo parameters ($N_{term}$, $N_{MCS}$ and $T_{sim}$) and results ($\Im m(x_c(L))$ and $T_c(L)$) for the 6 states Potts model. $N_{term}$ is the number of Monte Carlo steps used to equilibrate the system, $N_{MCS}$ stands for the number of measurements to build the histogram and $T_{sim}$ corresponds to each temperature in the iterative process.

| L | $N_{term}$ | $N_{MCS}$ | $T_{sim}$ | $\Im m(x_c(L))$ | $T_c(L)$ |
|---|---|---|---|---|---|
| 20 | $10^5$ | $5 \times 10^6$ | 0.8100 | 0.2118(7) | 0.814(2) |
| 40 | $10^5$ | $5 \times 10^6$ | 0.8100 | 0.669(3) | 0.8098(8) |
| 60 | $3 \times 10^5$ | $5 \times 10^6$ | 0.810 | | |
| | $3 \times 10^5$ | $5 \times 10^6$ | 0.809 | | |
| | $3 \times 10^5$ | $5 \times 10^6$ | 0.808 | 0.3387(8) | 0.80880(5) |
| 80 | $3 \times 10^5$ | $5 \times 10^6$ | 0.810 | | |
| | $3 \times 10^5$ | $5 \times 10^6$ | 0.809 | | |
| | $3 \times 10^5$ | $5 \times 10^6$ | 0.808 | 0.208(3) | 0.8085(5) |
| 120 | $3 \times 10^5$ | $5 \times 10^6$ | 0.810 | | |
| | $3 \times 10^5$ | $5 \times 10^6$ | 0.809 | | |
| | $3 \times 10^5$ | $5 \times 10^6$ | 0.808 | | |
| | $3 \times 10^5$ | $5 \times 10^6$ | 0.807 | 0.109(3) | 0.808(1) |

transition the scaling functions need to be considered in more detail [8]. In Figs. 4 and 5 are shown both the imaginary part of $x_c(L)$ and the pseudo-critical temperature as a function of $L^{-d}$. A linear adjust does not work in this case, anticipating that a correction is missing in the scaling function. Following Lee and Kostelitz [10] we introduce a $L^{-2d}$ scaling correction which fits quite well our results as seen in figs. 4 and 5. The critical temperature encountered (0.80797(7)) agrees up to 5 figures with the exact value (0.80761).

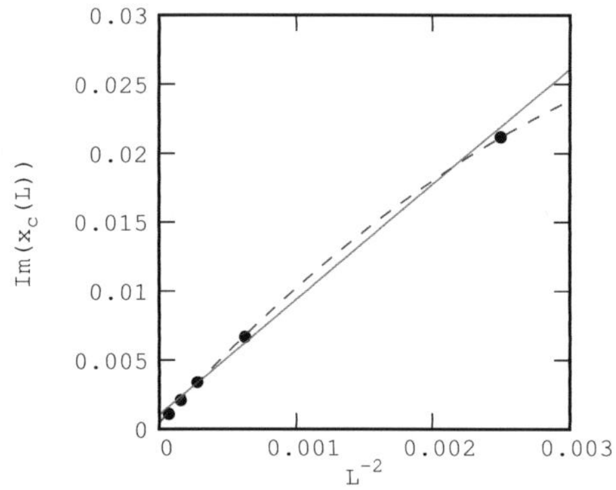


**Figure 4.** $\Im m(x_c(L))$ versus $L^{-2}$ for the 6 states Potts model. The solid red line is a linear adjust. The dashed blue line is a quadratic fit corresponding to scaling corrections.

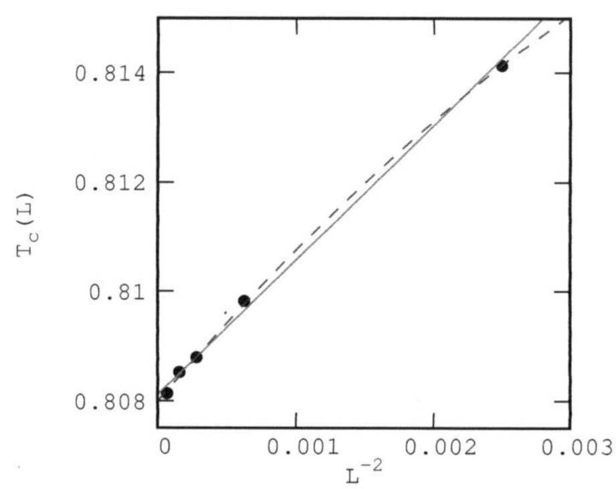


**Figure 5.** $T_c(L)$ versus $L^{-2}$ for the 6 states Potts model. Solid line is a linear fit. When the quadratic scaling correction is introduced (dashed blue line) the data are very well adjusted. From the adjust we get 0.80797(7) while the exact value is 0.80761.

## 5. Closing Remarks

In summary, we introduced a novel method to study phase transitions based on the zeros of the energy probability distribution. The method has shown to be fully applicable to the models studied allowing accurate determination of the critical temperatures and the exponent $\nu$ as demonstrated by our results in the Ising and Potts models. More general applications are found in the reference by Costa et al [11]. The only criterion we have used was the fact that at the transition temperature the partition function zero goes to the real positive axis in the thermodynamic limit. No *a priori* knowledge of an order parameter is necessary. The method is iterative, so that, the transition temperature can be approached at will. Besides that, the knowledge of the complete density of states is not required since we work in a restricted range of energy. The advantage is that the polynomial representing the partition function has fewer roots with its coefficients ranging in a civilized, narrow region.

**Acknowledgments**

This work was partially supported by CNPq and Fapemig, Brazilian Agencies. JCSR thanks CNPq for the support under grant PDJ No. 150503/2014-8 and Grant CNPq 402091/2012-4.

IOP Conf. Series: Journal of Physics: Conf. Series **921** (2017) 012004 doi:10.1088/1742-6596/921/1/012004

## References


[1] M. E. Fisher, Renormalization group theory: Its basis and formulation in statistical physics, Rev. Mod. Phys. 70 (1998) 653-681. doi:10.1103/RevModPhys.70.653.

[2] C. N. Yang, T. D. Lee, Statistical theory of equations of state and phase transitions. i. theory of condensation, Phys. Rev. 87 (1952) 404-409. doi:10.1103/PhysRev.87.404.

[3] M. E. Fisher, in: W. Brittin (Ed.), Lectures in Theoretical Physics: Volume VII C - Statistical Physics, Weak Interactions, Field Theory : Lectures Delivered at the Summer Institute for Theoretical Physics, University of Colorado, Boulder, 1964, no. v. 7, University of Colorado Press, Boulder, 1965.

[4] A. M. Ferrenberg, R. H. Swendsen, New Monte Carlo technique for studying phase transitions, Phys. Rev. Lett. 61 (1988) 2635-2638. doi:10.1103/PhysRevLett.61.2635.

[5] A. M. Ferrenberg, R. H. Swendsen, Optimized Monte Carlo data analysis, Phys. Rev. Lett. 63 (1989) 1195-1198. doi:10.1103/PhysRevLett.63.1195.

[6] L. Onsager, Crystal statistics. i. a two-dimensional model with an order-disorder transition, Phys. Rev. 65 (1944) 117-149. doi:10.1103/PhysRev.65.117.

[7] P. D. Beale, Exact distribution of energies in the two-dimensional Ising model, Phys. Rev. Lett. 76 (1996) 78-81. doi:10.1103/PhysRevLett.76.78.

[8] V. Privman, Finite Size Scaling and Numerical Simulation of Statistical Systems, World Scientific, 1990.

[9] F. Y. Wu, The Potts model, Rev. Mod. Phys. 54 (1982) 235-268. doi:10.1103/RevModPhys.54.235.

[10] J. Lee, J. M. Kosterlitz, Finite-size scaling and Monte Carlo simulations of First-order phase transitions, Phys. Rev. B 43 (1991) 3265-3277. doi:10.1103/PhysRevB.43.3265.

[11] B.V. Costa, L.A.S. Ml, J.C.S. Rocha, Energy Probability Distribution Zeros: A Route to Study Phase Transitions, Computer Physics Communications 216(2017)77. http://dx.doi.org/10.1016/j.cpc.2017.03.003.

IOP Conf. Series: Journal of Physics: Conf. Series **921** (2017) 012005 doi:10.1088/1742-6596/921/1/012005

# Applying Renormalization Group to Quantum Walks

**S. Boettcher**
Department of Physics, Emory University, Atlanta, GA 30322; USA

**Abstract.** We provide a brief introduction to quantum walks. In particular, we describe how to administer and interpret the real-space renormalization group for such walks, a subtle task due to replacing the stochastic propagator in classical random walks by a unitary operator.

## 1. Introduction

For random walks, the probability density $\rho(\vec{x},t)$ to detect a walk at time $t$ at site $\vec{x}$, a distance $x = |\vec{x}|$ from its origin, obeys the scaling collapse with the scaling variable $x/t^{1/d_w}$,

$$\rho(\vec{x},t) \sim t^{-\frac{d_f}{d_w}} f\left(x/t^{\frac{1}{d_w}}\right), \tag{1}$$

where $d_w$ is the walk-dimension and $d_f$ is the fractal dimension of the network [1]. On a translationally invariant lattice of any spatial dimension $d(= d_f)$, it is easy to show that the walk is always purely "diffusive", $d_w = 2$, with a Gaussian scaling function $f$, which is the content of many classic textbooks on random walks and diffusion [2]. The scaling in Eq. (1) still holds when translational invariance is broken or the network is fractal (i.e., $d_f$ is non-integer). Such "anomalous" diffusion with $d_w \neq 2$ may arise in many transport processes [1, 3]. For quantum walks, the only known value for a finite walk dimension is that for ordinary lattices [4], where Eq. (1) generically holds with $d_w = 1$, indicating a "ballistic" spreading of the quantum walk from its origin. This value has been obtained for various versions of one- and higher-dimensional quantum walks, for instance, with so-called weak-limit theorems [4, 5].

The renormalization group (RG) [6] expands the analytic tools to understand quantum walks [7, 8, 9, 10], since it works for networks that lack translational symmetries. We have obtained exact results for complex networks, with the scaling exponents shown in Tab. 1. We conjecture that the quantum walk dimension $d_w$ with a Grover coin always is half of that for the corresponding random walk [9],

$$d_w^{QW} = \frac{1}{2} d_w^{RW}. \tag{2}$$

This ability to explore a given geometry much faster than diffusion is essential for the effectiveness of quantum search algorithms [11, 12].

## 2. Evolution equation for a walk

Our walks are governed by the discrete-time "master" equation [3] for the state of the system,

$$|\Psi(t+1)\rangle = \mathcal{U}\,|\Psi(t)\rangle \tag{3}$$

IOP Conf. Series: Journal of Physics: Conf. Series **921** (2017) 012005 doi:10.1088/1742-6596/921/1/012005

**Table 1.** Walk dimensions for (hypercubic) lattices and some complex networks, as determined with the RG. The classical value for $d_w^{RW}$ is known for DSG [1] and HN3 [13], or easily derived for MK [9]. The values for $d_w^{QW}$ were first conjectured [8, 9], then analytically verified (except HN3) [10] for a reflective Grover coin. In each case, these $d_w$ satisfies Eq. (2).

| Network | $d_w^{RW}$ | $d_w^{QW}$ (Grover) |
|---|---|---|
| Lattice | 2 | 1 |
| MK3 | $\log_4(21) \approx 2.196$ | $1.098079\ldots$ |
| MK4 | $\log_4\left(\frac{247}{7}\right) \approx 2.571$ | $1.285253\ldots$ |
| HN3 | $\log_2\left(24 - 8\sqrt{5}\right) \approx 2.612$ | $1.305758\ldots$ |
| DSG | $\log_2(5) \approx 2.322$ | $1.160964\ldots$ |

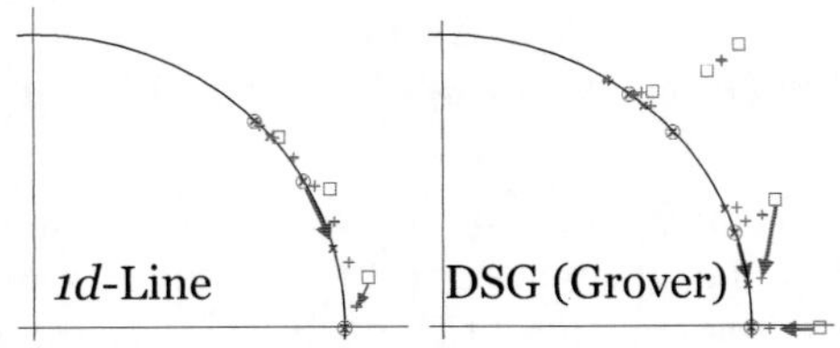


**Figure 1.** Plot of Laplace-poles in the complex-$z$ plane at RG-steps $k = 2, 3$ for site-amplitude $\overline{\psi}_0^{(k)}(z)$ (in blue, with $\circ$ for $k = 2$ and $\times$ for $k = 3$) and for the hopping parameters $\vec{a}_k(z)$ (in red, with $\Box$ for $k = 2$ and $+$ for $k = 3$) for quantum walk on the $1d$-line and DSG with a Grover coin. All poles for the site-amplitude are on the unit-circle, while all poles for the hopping parameters are outside. Either type of poles impinge increasingly on the real-$z$ axis (arrows).

with propagator $\mathcal{U}$. It is a stochastic operator for a classical, dissipative random walk, but in the quantum case $\mathcal{U}$ is unitary and, thus, reversible [yet, we will still refer to Eq. (3) as a master equation]. Then, in the discrete $N$-dimensional site-basis $|x\rangle$ of some network, the PDF is given by $\rho(x,t) = \psi_{x,t} = \langle x|\Psi(t)\rangle$ for random walks, or by $\rho(x,t) = |\psi_{x,t}|^2$ for quantum walks.

Assuming that we possess the eigensolution for the propagator, $\mathcal{U}\phi_j = u_j\phi_j$ with eigenvalues $u_j$ and an orthonormal set of eigenvectors $\phi_j(x)$, then the formal solution of Eq. (3) becomes $\psi_{x,t} = \sum_j a_j u_j^t \phi_j(x)$. For a stochastic $\mathcal{U}$, aside from the unique $(+1)$-eigenvalue of the stationary state, the remaining eigenvalues have $|u_j| < 1$, thus, according to Eq. (3), the dynamics is determined by $\rho(\vec{x},t) \sim e^{-t/\tau}$ for large times $t$ with $\tau = -1/\ln\max_j\{|u_j| < 1\}$. In turn, for unitary $\mathcal{U}$ all eigenvalues are uni-modular, $|u_j| = 1$ for all $j$, such that $u_j = e^{i\theta_j}$ with real $\theta_j$. A discrete Laplace-transform [3] of the site amplitudes $\overline{\psi}_x(z) = \sum_{t=0}^{\infty}\psi_{x,t}z^t$ has its poles – hence those for $\overline{\rho}(x,z)$ – located on the unit-circle in the complex-$z$ plane [10],

$$\overline{\rho}(x,z) \propto \prod_{j,l}\left[1 - z\,e^{i(\theta_j-\theta_l)}\right]^{-1}. \tag{4}$$

*Walks on the* 1d-*line:* To appreciate the promise of the RG, we review the time evolution of random and quantum walks on the *1d*-line, although these are well-understood and quantum search can not improve on the classical result here. The propagator in Eq. (3) is

$$\mathcal{U} = \sum_x \{A\,|x+1\rangle\langle x| + B\,|x-1\rangle\langle x| + M\,|x\rangle\langle x|\} \tag{5}$$

for nearest-neighbor transitions. For quantum walks the norm of $\rho$ demands unitary propagation, $\mathbb{I} = \mathcal{U}^\dagger\mathcal{U}$. The rules then impose the conditions $\mathbb{I}_r = A^\dagger A + B^\dagger B + M^\dagger M$ and $0 = A^\dagger M + M^\dagger B = A^\dagger B$, consistent with $A + B + M$ being unitary. This algebra requires at least $r = 2$-dimensional matrices, and it is customary [14] to use the most general unitary coin-matrix,

$$\mathcal{C} = \begin{pmatrix} \sin\eta & e^{i\chi}\cos\eta \\ e^{i\vartheta}\cos\eta & -e^{i(\chi+\vartheta)}\sin\eta \end{pmatrix}, \tag{6}$$

to define $\mathcal{U} = \mathcal{S}\left(\mathcal{C}\otimes\mathbb{I}_N\right)$ with shift $\mathcal{S}$ using matrices $S^{\{A,B,M\}}$ for transfer in a direction either out of a site or back to itself, i.e., $A = S^A\mathcal{C}$, $B = S^B\mathcal{C}$, and $M = S^M\mathcal{C}$ with $S^A + S^B + S^M = \mathbb{I}_r$. The quantum-coin entangles all $r$ components of $\psi_{x,t}$ and the shift-matrices facilitate the subsequent transitions to neighboring sites. For $r = 2$, the degree of each site, there are no self-loops ($S^M = 0, M = 0$) and we shift upper (lower) components of each $\psi_{x,t}$ to the right (left) using projectors $S^A = \begin{bmatrix} 1 & 0 \\ 0 & 0 \end{bmatrix}$ and $S^B = \begin{bmatrix} 0 & 0 \\ 0 & 1 \end{bmatrix}$. Alternatively, these coin-degrees of freedom could be replaced by more intricate next-nearest-neighbor hoppings without coin [15, 16], for which schemes equivalent to the following RG can be developed.

*Renormalization of walks:* To obtain the RG-recursions, a Laplace-transform is used such that the master equation (3) with $\mathcal{U}$ in Eq. (5) becomes $\overline{\psi}_x = zM\overline{\psi}_x + zA\overline{\psi}_{x-1} + zB\overline{\psi}_{x+1}$. We first consider the walk problem here with initial conditions (IC) localized at the origin, $\psi_{x,t=0} = \delta_{x,0}\psi_{IC}$. We recursively eliminating $\overline{\psi}_x$ for all sites for which $x$ is an odd number, then set $x \to x/2$ and repeat, step-by-step for $k = 0, 1, 2, \ldots$. Starting at $k = 0$ with the "raw" hopping coefficients $A_0 = zA$, $B_0 = zB$, and $M_0 = zM$, after each step, the master equation becomes *self-similar* in form when redefining the renormalized hopping coefficients $A_k$, $B_k$, $M_k$. E.g., for sites near any even site $x$ at step $k$ we have [7]:

$$\begin{aligned} \overline{\psi}_{x-1} &= M_k\overline{\psi}_{x-1} + A_k\overline{\psi}_{x-2} + B_k\overline{\psi}_x, \\ \overline{\psi}_x &= M_k\overline{\psi}_x + A_k\overline{\psi}_{x-1} + B_k\overline{\psi}_{x+1} + \delta_{x,0}\psi_{IC}, \\ \overline{\psi}_{x+1} &= M_k\overline{\psi}_{x+1} + A_k\overline{\psi}_x + B_k\overline{\psi}_{x+2}. \end{aligned} \tag{7}$$

Solving this *linear* system at $x$ yields $\overline{\psi}_x = M_{k+1}\overline{\psi}_x + A_{k+1}\overline{\psi}_{x-2} + B_{k+1}\overline{\psi}_{x+2}$ with RG "flow"

$$\begin{aligned} A_{k+1} &= A_k\left(\mathbb{I} - M_k\right)^{-1} A_k, \qquad B_{k+1} = B_k\left(\mathbb{I} - M_k\right)^{-1} B_k, \\ M_{k+1} &= M_k + A_k\left(\mathbb{I} - M_k\right)^{-1} B_k + B_k\left(\mathbb{I} - M_k\right)^{-1} A_k. \end{aligned} \tag{8}$$

For the quantum walk, we evolve Eqs. (8) from its unrenormalized values [7]. Already after two iterations, a recurring pattern emerges that suggest the Ansatz $A_k = a_k S^A \mathcal{C}$, $B_k = -a_k S^B \mathcal{C}$, and $M_k = m_k \begin{bmatrix} 0 & 1 \\ 1 & 0 \end{bmatrix} \mathcal{C}$. For $a_k$ and $m_k$, the RG-flow (8) closes for

$$a_{k+1} = \frac{a_k^2 \sin\eta}{1 - 2m_k\cos\eta + m_k^2}, \qquad m_{k+1} = m_k + \frac{(m_k - \cos\eta)\, a_k^2}{1 - 2m_k\cos\eta + m_k^2} \tag{9}$$

for $0 < \eta < \pi/2$ and $\chi = \vartheta = 0$ in Eq. (6). These recursions have a non-trivial fixed point at $(a_\infty, m_\infty) = (\sin\eta, \cos\eta)$, yet, its Jacobian at $k \to \infty$ is, in fact, $\eta$-independent and has a degenerate eigenvalue $\lambda_{1,2} = 2$, suggesting $d_w = \log_2 \lambda_1 = 1$. Basically, the same procedure as in Sec. 2, although algebraically more laborious, can be applied to obtain closed RG-recursions for a number of fractal networks [8, 9]. These have provided the first exact weak-limit results for non-trivial networks beyond regular lattices, as listed in Tab. 1.

## 3. Interpreting the RG

Although the above analysis of the fixed points of the RG-flow appear to follow the conventional approach [3] for both, the classical and the quantum walk, puzzling discrepancies emerge when such a naive approach is extended beyond *1d*. When applied to the dual Sierpinski gasket (DSG) [8], or a number of other hierarchical networks [9], the largest Jacobian eigenvalue corresponds merely to their fractal dimension, i.e., $\log_2 \lambda_1 = d_f$. It is an accident that $d_f = d_w^{QW} = 1$ on the *1d*-line! Hence, we had to reconsider the basis of the Tauberian theorems [2, 3] and obtained the following interpretation [10]: The poles closest to $z_\infty = 1$ in the

Laplace-transforms $\overline{\psi}_x(z)$ determine the large-$t$ behavior of the walk dynamics. For increasing length (i.e., $k \to k+1$ in the RG-flow), these poles move towards $z_\infty$ in a characteristic way [3]. In the classical case, for a stochastic propagator in Eq. (3), these poles move radially along the real-$z$ axis with $z_k > 1$ and only the pole with smallest to $|z_k/z_\infty| - 1$ determines the scaling between length and time in Eq. (1). In the quantum case, with a unitary operator, *all* poles of $\overline{\psi}_x(z)$ are unimodular, i.e., $|z_k/z_\infty| \equiv 1$ for all $k$, see Eq. (4). Instead, they approach $z_\infty$ *tangentially* along the complex unit-circle in a characteristic manner that provides the correct scaling. Yet, this behavior is somewhat obscured within the RG-flow: The poles of hopping parameters, such as $a_k(z)$ or $m_k(z)$ in Eq. (9), for example, move radially as well as tangentially, see Fig. 1. As in the classical case, it is the radial motion of poles that is captured by the largest Jacobian eigenvalue $\lambda_1$, while we have shown [10] that the tangential motion is described by the *geometric mean* of the *two* largest eigenvalues $\lambda_{1,2}(>1)$, so that $d_w^{QW} = \log_2\sqrt{\lambda_1\lambda_2}$ indeed reproduces the correct quantum scaling consistent with the numerical results (to within some 6-8 digits accuracy!) [8, 9]. For example, the RG for quantum walk on DSG yields $\lambda_1 = 3$, $\lambda_2 = \frac{5}{3}$, such that $\sqrt{\lambda_1\lambda_2} = \sqrt{5}$ reproduces the numerically predicted scaling, $d_w^{QW} = \log_2\sqrt{5}$. Since classically $\lambda_1^{\rm RW} = 5 (= \lambda_1\lambda_2)$, it verifies the conjecture in Eq. (2). The same pattern holds for two Migdal-Kadanoff networks [9]: For the 3-regular network called MK3, the Jacobian eigenvalues are $\lambda_1 = 7$ and $\lambda_2 = 3$ such that $\lambda_1\lambda_2 = \lambda_1^{\rm RW} = 21$, and for the 4-regular network called MK4 they are $\lambda_1 = 13$ and $\lambda_2 = \frac{19}{7}$, such that $\lambda_1\lambda_2 = \lambda_1^{\rm RW} = \frac{247}{7}$.

## 4. Conclusions

Some remarks are in order: First, it is unfortunate that the quantum walk on the *1d*-line, the only independently solvable model to compare to the RG, provides no insights because $\lambda_1 = \lambda_2 = \sqrt{\lambda_1\lambda_2} = 2$ can not distinguish between radial and tangential motion of poles. Second, our explanation of the quantum case here is somewhat simplified because the poles of observables like $\rho(x,t) = |\psi_{x,t}|^2$ depend on $|\psi_{x,t}|^2$ and thus those poles result from all mutual *ratios* $\exp\{i(\theta_n - \theta_m)\}$ of unimodular eigenvalues $\exp\{i\theta_j\}$ of the unitary propagator $\mathcal{U}$, see Eq. (4). We argue [10] that, if *all* such eigenvalues scale with $k$ as $\theta_j^{(k)} \sim (\lambda_1\lambda_2)^{-\frac{k}{2}}$ or as a constant, then so does any difference $\theta_n - \theta_m$, and thus it is sufficient to follow any convenient pole of $a_k(z)$ such as that nearest to $z_\infty = 1$. Unfortunately, this assumption appears to fail in some cases, like the network HN3 we considered before [9], see Tab. 1.

SB is supported by the NSF through grant DMR-1207431.

## References

[1] Havlin S and Ben-Avraham D 1987 *Adv. Phys.* **36** 695–798
[2] Weiss G H 1994 *Aspects and Applications of the Random Walk* (North-Holland, Amsterdam)
[3] Redner S 2001 *A Guide to First-Passage Processes* (Cambridge: Cambridge University Press)
[4] Grimmett G, Janson S and Scudo P F 2004 *Physical Review E* **69** 026119+
[5] Konno N 2002 *Quantum Information Processing* **1** 345–354
[6] Pathria R K 1996 *Statistical Mechanics, 2nd Ed.* (Butterworth-Heinemann, Boston)
[7] Boettcher S, Falkner S and Portugal R 2013 *Journal of Physics: Conference Series* **473** 012018
[8] Boettcher S, Falkner S and Portugal R 2014 *Phys. Rev. A* **90** 032324
[9] Boettcher S, Falkner S and Portugal R 2015 *Phys. Rev. A* **91** 052330
[10] Boettcher S, Li S and Portugal R 2017 *J. Phys. A* **50** 125302
[11] Ambainis A 2007 *SIAM J. Comput.* **37** 210–239
[12] Childs A M, Cleve R, Deotto E, Farhi E, Gutmann S and Spielman D A 2003 *Proceedings of the Thirty-fifth Annual ACM Symposium on Theory of Computing* STOC '03 (New York, NY, USA: ACM) pp 59–68
[13] Boettcher S and Gonçalves B 2008 *Europhysics Letters* **84** 30002
[14] Portugal R 2013 *Quantum Walks and Search Algorithms* (Springer, Berlin)
[15] Portugal R, Boettcher S and Falkner S 2015 *Phys. Rev. A* **91** 052319
[16] Portugal R 2016 *Quantum Information Processing* **15** 1387–1409

IOP Conf. Series: Journal of Physics: Conf. Series **921** (2017) 012006 doi:10.1088/1742-6596/921/1/012006

# Collective motion in repulsive self-propelled particles in confined geometries

**Takayuki Hiraoka[1,†], Takashi Shimada[1,2] and Nobuyasu Ito[1,3]**

[1] Department of Applied Physics, The University of Tokyo, Tokyo 113-8656, Japan
[2] Quantum-Phase Electronics Center, School of Engineering, The University of Tokyo, Tokyo 113-8656, Japan
[3] RIKEN Advanced Institute for Computational Science, Kobe, Hyogo 650-0047, Japan

E-mail: hiraoka@serow.t.u-tokyo.ac.jp

**Abstract.** We study the collective dynamics of a simple model of repulsive self-propelled particles in three different geometries bounded by reflecting walls. Order-disorder transition is observed in all three cases and the behavior of the phase diagram is similar to that of a system with periodic boundary conditions, which supports a view that the coherent motion of particles is induced by a microscopic mechanism. We also find that some interesting quasi-stable dynamic patterns may appear during the ordering process. The long lifetime of the transient structures indicates that the boundaries hinder the propagation of correlations.

## 1. Introduction

Much efforts have been devoted to understand the universal feature of collective behaviors seen in biological systems over the last two decades. One of the earliest computational models is the Vicsek model [1], which assumes ferromagnetic-like alignment interactions between self-propelled point particles. Despite its minimal feature, the model exhibits characteristic dynamical patterns, such as a spontaneous transition to collective motion, a long wavelength instability at the onset of order, and so-called "giant" density fluctuations [2, 3]. Alignment with neighbors, however, is not always the origin for the collective behavior empirically observed. In fact, many experimental results suggest that the repulsion and excluded volume play an important role in bacterial colonies and cellular aggregates [4–6]. Pedestrian movement in human crowds is also influenced by socio-psychological repulsion from others [7,8]. Collective dynamics with pure repulsion or hard steric interaction (without alignment) has been investigated for quite some time [9–13]. However, only a few studies [14, 15] have explored the full phase diagram with a focus on ordering behavior. In our previous work [15], we argued that repulsive active particles may undergo a transition from disorder to coherent motion, based on the results from numerical simulations with periodic boundary conditions (PBCs).

The PBCs gain a long-standing popularity in the field of computational statistical physics because they are useful for approximating the bulk properties from finite size simulations. However, they have a limitation that an artificial order may be imposed that can lead to abnormal results for structural and dynamical properties. In equilibrium systems, finite size scaling analysis provides a convenient way to infer the asymptotic behavior in the thermodynamic limit. However, the method may neither be justifiable nor come in handy when applied to systems strongly out of equilibrium. In such systems, the effect of geometry and of

† Present address: Asia Pacific Center for Theoretical Physics, Pohang, Gyeongsangbuk-do 37673, Republic of Korea.

boundaries on the dynamics must be examined by direct simulations. In the context of self-propelled particles, some geometries are particularly important because of their analogy with physical systems. In the embryonic development process, for example, collective cell migration often takes place in certain spatial confinements surrounded by extracellular matrix [16, 17]; similarly, pedestrian movement responds to spatial structure of the environment, which is defined by walls and obstacles [18]. In this paper, we analyze the dynamics of the self-propelled particles in systems with three basic geometries: A *pipe* or a *corridor* has periodic boundaries in the $x$ direction while bounded by two reflecting walls in the $y$ direction; a *box* or a *room* is a rectangular area surrounded by walls at all four sides; a *disk* is a circular surface also enclosed by a reflecting boundary.

## 2. Repulsive self-propelled particle model

We consider disk particles $i$ of radius $a$, driven by self-propelling force of constant magnitude $\alpha$ in the direction of polarity $\psi_i$. The dynamics is damped both in the translational and rotational degrees of freedom: a particle is subject to a friction proportional to the velocity with coefficient $\beta$, and a torque which is proportional to the angular difference between the velocity and the polarity with coefficient $\gamma$, which rotates the polarity. The overall dynamics is described by the following equations of motion:

$$\frac{\mathrm{d}\mathbf{v}_i}{\mathrm{d}t} = \alpha\hat{\mathbf{e}}(\psi_i) - \beta\mathbf{v}_i + \sum_{j\neq i}\mathbf{f}_{ij} + \sum_{W}\mathbf{f}_{iW}, \tag{1}$$

$$\frac{\mathrm{d}\psi_i}{\mathrm{d}t} = \gamma(\theta_i - \psi_i). \tag{2}$$

Here, $\hat{\mathbf{e}}(\cdot)$ indicates an unit vector in the direction of the argument, and $\theta_i$ is the direction of the velocity. The interaction with another particle $j$ is a Hookean contact force,

$$\mathbf{f}_{ij} = \begin{cases} -k\left(2a - |\mathbf{r}_{ij}|\right)\dfrac{\mathbf{r}_{ij}}{|\mathbf{r}_{ij}|} & (|\mathbf{r}_{ij}| < 2a) \\ \mathbf{0} & (\text{otherwise}), \end{cases} \tag{3}$$

where $\mathbf{r}_{ij} = \mathbf{r}_i - \mathbf{r}_j$. We assume a similar interaction with a reflecting boundary $W$

$$\mathbf{f}_{iW} = \begin{cases} -k\left(a - |\mathbf{r}_{iW}|\right)\dfrac{\mathbf{r}_{iW}}{|\mathbf{r}_{iW}|} & (|\mathbf{r}_{iW}| < a) \\ \mathbf{0} & (\text{otherwise}), \end{cases} \tag{4}$$

where $\mathbf{r}_{iW}$ denotes the normal vector from the position of the particle to the boundary. The stiffness $k$ is constant and the same for both types of interactions.

In our previous work [15], we fixed values of $\alpha$, $\beta$, and $k$ so the only control parameter left for microscopic dynamics will be the rotational damping coefficient $\gamma$, and performed simulations in a square box with PBCs. An ordered motion spontaneously emerges from an initially random configuration for small values of $\gamma$. The system tends to remain longer in an isotropic and disordered state before the transition for larger $\gamma$; the waiting time eventually exceeds the finite simulation time. We showed that this behavior stems from iteration of binary collision between particles. In this paper, we examine the dynamics with different geometries and boundary conditions.

## 3. *Pipe* geometry

First, we discuss the behavior of the system in a *pipe*, of which the vertical boundaries (perpendicular to the $x$ axis) are periodic while the horizontal ones (perpendicular to the $y$ axis) are reflecting. Typical snapshots are shown in the left column of Fig. 1. The most significant difference from the full-periodic case is the aggregation along the walls. A configuration with

IOP Conf. Series: Journal of Physics: Conf. Series **921** (2017) 012006 doi:10.1088/1742-6596/921/1/012006

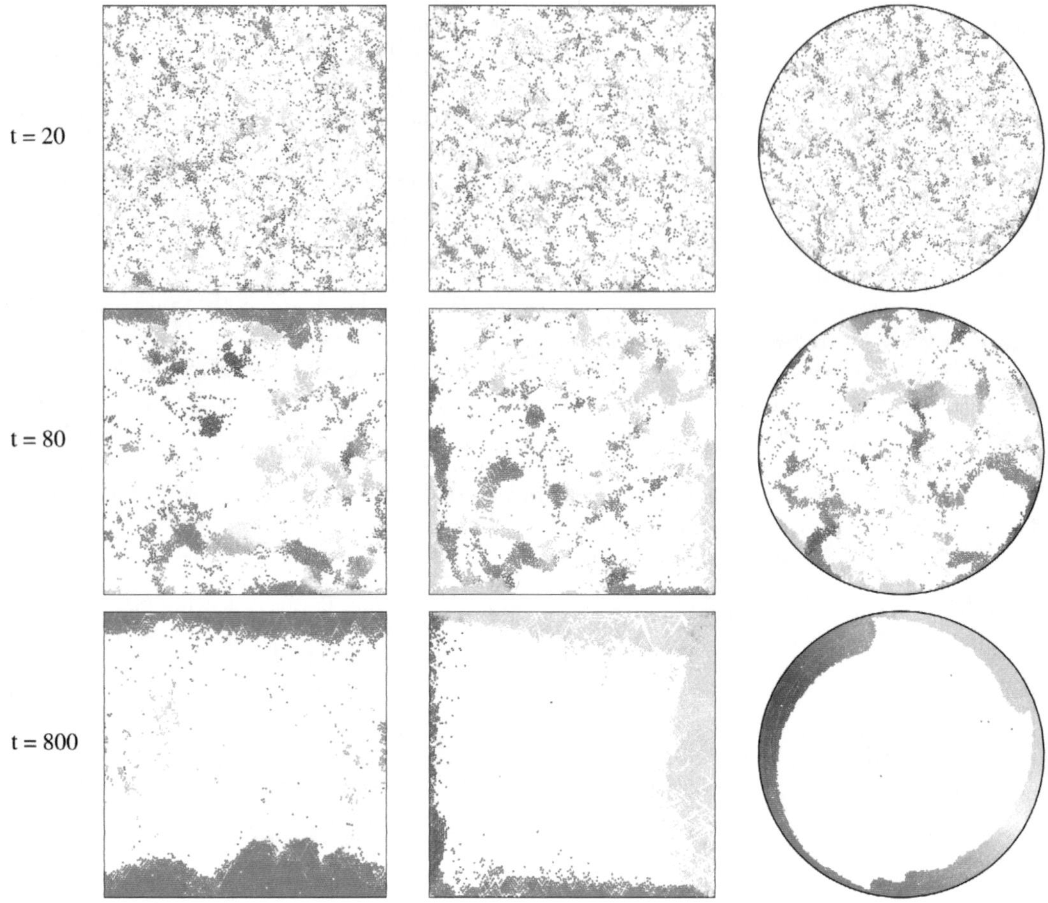


Figure 1: Snapshots of repulsive self-propelled particles at successive times in three different geometries. Colors represent the direction of the polarity. From left to right: time development in a *pipe*, in a *box*, and in a *disk*, respectively. The parameters are $N = 5000$, $\Phi = 0.2$ and $\gamma = 1$ for all figures.

two streams moving in the opposite directions at the reflecting borders is stable, and a fully ordered motion over the entire system cannot be achieved once this structure is formed. A single stream or multiple lanes, moving parallel to the $x$ axis, are also stable. The directions of streams and the number distribution of the aggregates do not depend on the parameter values.

In order to quantify the degree of coherence, we adopt an order parameter following the previous studies on lane formation of colloidal particles [19]. First, we classify the particles into two groups, those who are pointing in the positive direction and those pointing in negative direction, based on the sign of $\hat{\mathbf{e}}(\psi_i) \cdot \hat{\mathbf{x}}$. Then we assign each of them an order parameter $\phi_i^{\text{pipe}}$, which is equal to 1 if the distance along the $y$ axis, $|y_i - y_j|$, to all the particles $j$ in the other group is larger than the sum of the radius of the two particles and equal to 0 otherwise. Finally, the global order parameter $\phi^{\text{pipe}}$ is defined as the fraction of "ordered" particles,

$$\phi^{\text{pipe}} = \frac{1}{N} \sum_{i=1}^{N} \phi_i^{\text{pipe}}. \tag{5}$$

The phase diagram as a function of volume fraction $\Phi$ and $\gamma$, as shown in Fig. 2(a), is generated by averaging the values of the order parameter for the first 5000 time steps. It is in good agreement with the phase diagram for the full-periodic case in Ref. [15].

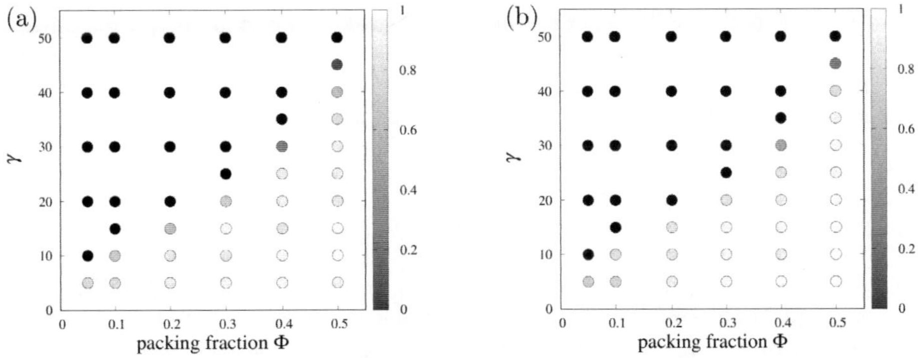


Figure 2: Phase diagrams for $N = 5000$ system (a) with the *pipe* geometry, and (b) with the *disk* geometry. The dots are shaded from black to white proportional to the time average of the order parameter over 5000 time units.

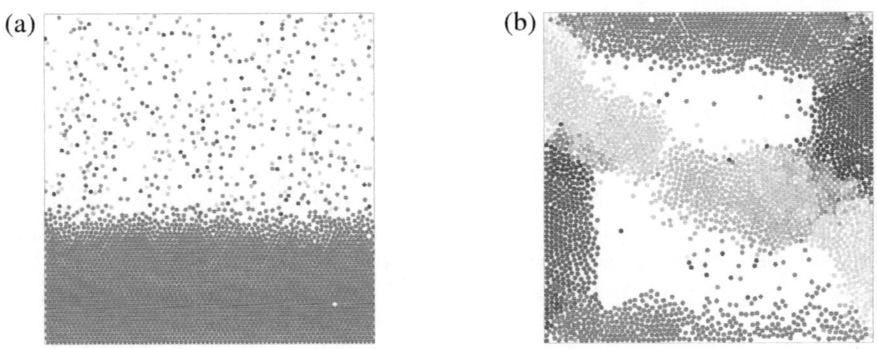


Figure 3: Snapshots of transient states. (a) The coexistence of "gaseous" state at the top and "solid" state at the bottom in a *pipe* The parameters are $N = 3000$, $\Phi = 0.4$, and $\gamma = 30$. (b) A transient state with two vortices, observed in a *box* confinement. $N = 3000$, $\Phi = 0.5$, and $\gamma = 10$.

In some runs near the phase boundary, we observe a transient state where particles are ordered and densely packed in one part while disordered in the rest of the system, as shown in Fig. 3(a). This coexistence between a "crystallized" region and a "gaseous" region is not a stationary state; particles in the gaseous region are eventually absorbed into the ordered stream. However, the relaxation tends to be much slower than that in a full-periodic system, where nucleation is followed by quick growth of clusters which results in the global polarization. We suppose this is due to the difference between the two geometries in capability of propagating the particle-particle correlations. The reflecting boundaries prevent the correlations to spread from one side to another.

## 4. Enclosed geometries: *box* and *disk*

Next we turn to the *box* geometry, where the reflecting walls are placed at all four sides of the simulation box. In contrast to the periodic and pipe cases, the entire system cannot move in a coherent direction because there is no inflow nor outflow. The "ordered" state in this geometry is a whirling motion along the boundary, as shown in the center column of Fig. 1. A state with two vortices, rotating in the direction opposite to each other, is a quasi-stable transient state (Fig. 3(b)). We emphasize that although this structure is not stationary it survives for a considerably long time compared to the rapid ordering in a system with PBCs. This is again supposed to be due to the bounded propagation of correlations.

Although the transition to coherent motion is evident from the snapshots, a suitable order parameter is difficult to construct for the box geometry. Instead of a rectangular box, we resort to another, more symmetric confinement: a simulation area enclosed by a circular reflecting wall. As one would expect, coherent circular motion is observed as an ordered state under this geometrical condition as well (Fig. 1, right column). We detect the transition to the circular flow by monitoring an order parameter $\phi^{\text{disk}}$, which is defined as

$$\phi^{\text{disk}} = \frac{1}{N}\left|\sum_{i=1}^{N}\frac{\mathbf{v}_i\cdot\hat{\mathbf{T}}_{\omega_i}}{|\mathbf{v}_i|}\right|, \tag{6}$$

where $\hat{\mathbf{T}}_{\omega_i} = -\sin\omega_i\hat{\mathbf{x}} + \cos\omega_i\hat{\mathbf{y}}$ is an unit tangent vector at the polar angle $\omega_i$ whose pole is located at the center of the disk. $\phi^{\text{disk}} = 1$ is attained when all the particles rotate along the concentric circles in the same direction. Again, we perform simulations for various pairs of parameter $(\Phi, \gamma)$ and depict the phase diagram as shown in Fig. 2(b). The fact that the phase boundary traces almost the same curve as those in the periodic and pipe cases leads to a conclusion that the phase transition is not affected by the change of boundary conditions and that the intrinsic mechanism behind the ordering behavior is indeed the correlations propagated by iteration of binary collisions.

## 5. Summary

We investigate behaviors of simple repulsive self-propelled particle systems with three different geometries, namely, the *pipe*, the *box*, and the *disk*. Since the microscopic mechanism leading to the coherent motion is universal regardless of the geometry, we can predict the behavior of the phase diagrams from that of the system with periodic boundary conditions. However, transient states, where particles are locally aligned but the entire system has not yet reached a stationary state, tend to have longer lifetimes in the systems bounded by reflecting walls, because they prevent the velocity correlations from being transmitted. These transient states include coexistence between gaseous and crystallized phase and convective patterns with multiple vortices. The numerical results we present here illustrate analogies as well as differences between experimental active matter systems, such as bacteria, in which the localization along the border is rarely observed. Bridging this gap would be an interesting future direction.

## Acknowledgements
This work was supported by CREST, JST.

## References

[1] Vicsek T, Czirók A, Ben-Jacob E, Cohen I and Shochet O 1995 *Phys. Rev. Lett.* **75** 1226–1229
[2] Grégoire G and Chaté H 2004 *Phys. Rev. Lett.* **92** 025702
[3] Chaté H, Ginelli F and Raynaud F 2008 *Phys. Rev. E* **77** 046113
[4] Szabó A, Ünnep R, Méhes E, Twal W O, Argraves W S, Cao Y and Czirók A 2010 *Phys. Biol.* **7** 046007
[5] Zhang H P, Be'er A, Florin E L and Swinney H L 2010 *Proc. Natl. Acad. Sci.* **107** 13626–13630
[6] Starruß J, Peruani F, Jakovljevic V, Søgaard-Andersen L, Deutsch A and Bär M 2012 *Interface Focus* **2** 774–785
[7] Helbing D and Molnár P 1995 *Phys. Rev. E* **51** 4282–4286
[8] Karamouzas I, Skinner B and Guy S J 2014 *Phys. Rev. Lett.* **113** 238701
[9] Peruani F, Deutsch A and Bär M 2006 *Phys. Rev. E* **74** 030904
[10] Grossman D, Aranson I S and Ben Jacob E 2008 *New J. Phys.* **10** 023036
[11] Henkes S, Fily Y and Marchetti M C 2011 *Phys. Rev. E* **84** 040301
[12] Fily Y and Marchetti M C 2012 *Phys. Rev. Lett.* **108** 235702
[13] Bialké J, Löwen H and Speck T 2013 *EPL* **103** 30008
[14] Hanke T, Weber C A and Frey E 2013 *Phys. Rev. E* **88** 052309
[15] Hiraoka T, Shimada T and Ito N 2016 *Phys. Rev. E* **94** 062612
[16] Haas P and Gilmour D 2006 *Dev. Cell* **10** 673–680
[17] Cetera M, Ramirez-San Juan G R, Oakes P W, Lewellyn L, Fairchild M J, Tanentzapf G, Gardel M L and Horne-Badovinac S 2014 *Nat. Commun.* **5** 5511
[18] Helbing D, Buzna L, Johansson A and Werner T 2005 *Transp. Sci.* **39** 1–24
[19] Dzubiella J, Hoffmann G P and Löwen H 2002 *Phys. Rev. E* **65** 021402

IOP Conf. Series: Journal of Physics: Conf. Series **921** (2017) 012007 doi:10.1088/1742-6596/921/1/012007

# Combined molecular and spin dynamics simulation of bcc iron with lattice vacancies

**Mark Mudrick[1], Markus Eisenbach[2], Dilina Perera[3], G. Malcolm Stocks[2] and David P. Landau[1]**

[1] Center for Simulational Physics, The University of Georgia, Athens , GA 30602, USA
[2] Oak Ridge National Laboratory, Oak Ridge, TN 37831, USA
[3] Department of Physics and Astronomy, Mississippi State University, Mississippi State, Mississippi 39762, USA

E-mail: mmudrick@uga.edu

**Abstract.** Using an atomistic model that treats both translational and spin degrees of freedom, we have performed combined molecular and spin dynamics simulations to study dynamic properties of BCC iron with varying vacancy concentrations. Atomic interactions are described by an empirical many-body potential while spin interactions are handled with a Heisenberg-like Hamiltonian with a coordinate dependent exchange interaction. By calculating the Fourier transform of spatial and temporal correlation functions, vibrational and magnetic excitations have been studied. The creation of vacancies in the material has shown splitting of the characteristic transverse spin-wave excitations, indicating the production of additional excitation modes. By merging two vacancies to form a nearest neighbor pair, we find that these modes become more distinct. Investigation of longitudinal spin-wave excitations reveals interactions between constituent components of the split transverse excitations.

## 1. Introduction

Computer simulation methods have been utilized for decades to study dynamic properties of materials at finite temperature. The molecular dynamics (MD) method [1, 2] has been widely used to investigate structural properties including diffusion, dislocations, and surface phenomena. To investigate time-dependent properties of magnetic systems, the spin dynamics (SD) method [3, 4] has given insight into propagating spin waves [5] and longitudinal two-spin-wave modes. However the coupling of lattice vibrations with spin waves has only been investigated in recent years [6], revealing a strong interplay [7] between the two subsystems. Therefore, models attempting to describe a real magnetic material must handle both the lattice and spin subsystems on equal footing. Recent work using the combined molecular and spin dynamics (MDSD) framework has only investigated these systems for pure conditions, while inpurities such as vacancies and interstitial defects are common in metallic materials. Vacancy defects are of importance in nuclear energy production, as radiation damage and hydrogen-vacancy interactions [8] may cause a metal to weaken over time.

## 2. Model and methods

### *2.1. Hamiltonian and equations of motion*

For a system of $N$ atoms of mass $m$ with positions $\{\mathbf{r}_i\}$, velocities $\{\mathbf{v}_i\}$ and classical atomic spins $\{\mathbf{S}_i\}$, the MDSD Hamiltonian is written

$$\mathcal{H} = \sum_{i=1}^{N} \frac{mv_i^2}{2} + U(\{\mathbf{r}_i\}) - \sum_{i<j} \mathbf{J}_{ij}(\{\mathbf{r}_k\})\mathbf{S}_i \cdot \mathbf{S}_j, \tag{1}$$

where the first term represents the kinetic energy of the atoms, and $U(\{\mathbf{r}_i\})$ is a many-body interatomic potential. In this work, the potential used is an embedded atom potential parameterized for iron by Dudarev and Derlet [9]. The third term determines the exchange coupling between the spins, and is Heisenberg-like with a coordinate-dependent exchange interaction $J_{ij}(\{\mathbf{r}_k\})$ parameterized for iron by first-principles calculations [10]. In the case of a vacancy at site $i$, the atom corresponding to the position of site $i$ is removed from the system. This Hamiltonian has true dynamics, described by the classical equations of motion

$$\begin{aligned} \frac{d\mathbf{r}_i}{dt} &= \mathbf{v}_i \\ \frac{d\mathbf{v}_i}{dt} &= \frac{\mathbf{f}_i}{m} \\ \frac{d\mathbf{S}_i}{dt} &= \frac{1}{\hbar}\mathbf{H}_i^{eff} \times \mathbf{S}_i, \end{aligned} \tag{2}$$

where $\mathbf{f_i} = -\nabla_{\mathbf{r}_i}\mathcal{H}$ and $\mathbf{H}_i^{eff} = \nabla_{\mathbf{S}_i}\mathcal{H}$ represent the interatomic force and effective field felt by atom $i$, respectively.

### *2.2. Integration scheme*

The time evolution of phase variables is written as

$$\frac{dq(t)}{dt} = \hat{L}q(t), \tag{3}$$

where $q(t) = \{\mathbf{r}_i(t), \mathbf{v}_i(t), \mathbf{e}_i(t)\}$ represents the complete set of positions, velocities, and spin orientations at time $t$. The Liouville operator $\hat{L}$ is of the form

$$\hat{L} = \sum_{i=1}^{N} \left( \mathbf{v}_i \cdot \frac{\partial}{\partial \mathbf{r}_i} + \frac{\mathbf{f}_i}{m_i} \cdot \frac{\partial}{\partial \mathbf{v}_i} + (\frac{g\mu_B}{\hbar M})(\mathbf{H}_i^{eff} \times \mathbf{e_i}) \cdot \frac{\partial}{\partial \mathbf{e}_i} \right). \tag{4}$$

The solution to equation 3 is given by

$$q(t+\tau) = e^{\hat{L}\tau} q(t), \tag{5}$$

where $\tau$ represents the finite time step and $e^{\hat{L}\tau}$ is the operator which produces the time evolution of the phase variables. Utilization of a second-order Suzuki-Trotter decomposition [11] allows for the decomposition of the exponential operator in Eq. (5). The equations in Eq. (2) can then be solved numerically, producing trajectories for the atomic and spin degrees of freedom.

### *2.3. Dynamic structure factor*

During the simulation, the microscopic spin density is calculated and the spin-spin intermediate scattering function is computed after each time step,

$$F^k_{ss}(\mathbf{q},t) = \frac{1}{N}\langle \rho^k_s(\mathbf{q},t)\rho^k_s(-\mathbf{q},0)\rangle, \tag{6}$$

where $k = x, y, z$ and $\rho^k_s(\mathbf{q},t)$ is the $k$-component of the microscopic spin density. Using the choice of a Cartesian coordinate system such that the $z$-axis is parallel to the magnetization vector, spin excitations that propagate parallel and perpendicular to this symmetry axis are differentiated. The components of $F_{ss}(\mathbf{q},t)$ are regrouped into a longitudinal component

$$F^L_{ss}(\mathbf{q},t) = F^z_{ss}(\mathbf{q},t), \tag{7}$$

and a transverse component

$$F^T_{ss}(\mathbf{q},t) = \frac{1}{2}(F^x_{ss}(\mathbf{q},t) + F^y_{ss}(\mathbf{q},t)). \tag{8}$$

The temporal Fourier transform of $F^{L,T}_{ss}(\mathbf{q},t)$ yields the dynamic structure factor,

$$S^{L,T}_{ss}(\mathbf{q},\omega) = \frac{1}{2\pi}\int_{-\infty}^{\infty} F_{ss}(\mathbf{q},t)e^{-i\omega t}dt, \tag{9}$$

for momentum transfer $\mathbf{q}$ and frequency $\omega$, an experimentally accessible quantity [12]. In each of the included figures, data have been smoothed for clarity using a moving average.

## 3. Results

### 3.1. Randomly distributed vacancies

To investigate the effects of randomly distributed vacancies, MDSD simulations have been performed on L=8 simulation cells (1024 lattice sites) with 5% of lattice sites left vacant and compared to results from pure lattice simulations. At a temperature of $T = 300$K, pure lattice results display a well defined single excitation peak [13] in the transverse component of the spin-spin dynamic structure factor $S^T_{ss}(\mathbf{q},\omega)$. The introduction of vacancies distributed throughout the lattice causes a dampening of transverse dynamic structure factor peaks as well as a shift to lower frequency. For small values of $|\mathbf{q}|$, the peak in the dynamic structure factor is shown to exhibit splitting into multiple peaks as seen in Fig.1.

### 3.2. Vacancy pair formation

The effect of merging vacant sites from distant positions outside of the interaction cut-off into nearest neighbor lattice positions in a small system is shown in Fig. 2. Removing two atoms from distant sites results in a dampening of the characteristic spin wave peak and a broadening toward lower frequency. As these vacancies are brought closer together, the single peak becomes split into two distinct peaks, located at frequencies higher and lower than the pure lattice excitation. This splitting behavior is most pronounced when the vacancies occupy nearest neighbor lattice positions.

Excitations in the longitudinal component of the dynamic structure factor $S^L_{ss}(\mathbf{q},\omega)$ represent two-spin-wave annihilation processes [14]. These excitations are typically much lower in intensity than transverse peaks. Frequencies of longitudinal excitations can be predicted by subtracting the frequencies of constituent transverse spin waves. In the case of a nearest neighbor vacancy pair, the longitudinal spin wave spectrum displays more excitations due to the increased number of transverse spin waves as indicated by the peak splitting behavior shown in Fig. 2. These new excitations can be identified as the interaction between constituent components of the split transverse peaks, and their locations are predicted as shown in Fig. 3.

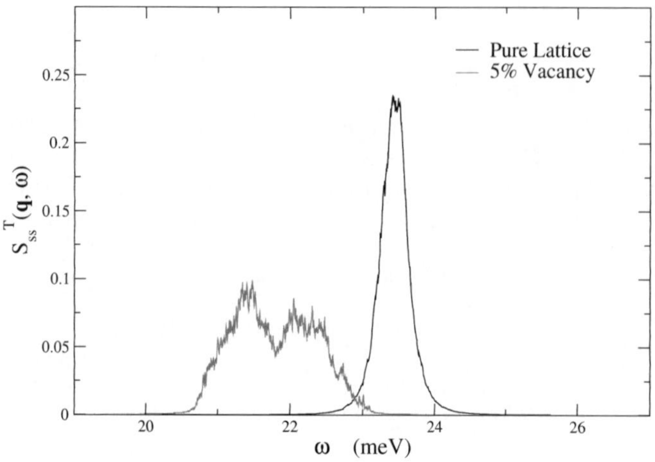


**Figure 1.** Transverse component of the spin-spin dynamic structure factor $S^T_{ss}(\mathbf{q}, \omega)$ in MDSD simulation at temperature $T = 300$K for $|\mathbf{q}| = 0.274\text{Å}^{-1}$ along the $(q, 0, 0)$ direction for an $L = 8$ simulation cell

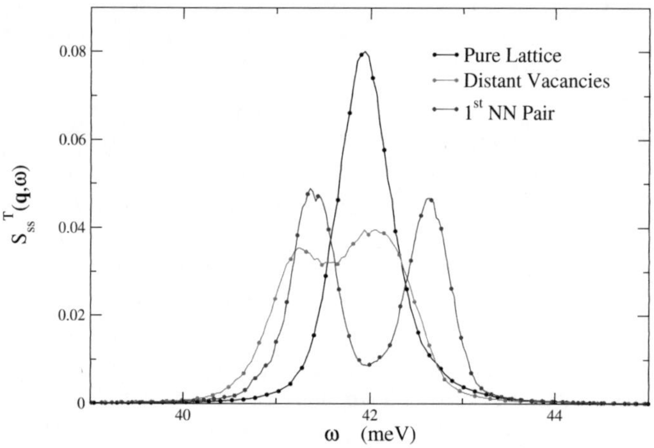


**Figure 2.** Transverse spin-spin dynamic structure factor for $|\mathbf{q}| = 0.365\text{Å}^{-1}$ along the $(q, 0, 0)$ direction in an L=6 system showing the merging of two defects into a nearest neighbor vacancy pair.

## 4. Summary

Using the combined molecular and spin dynamics simulation method, we have studied the effects of lattice vacancies on spin waves in BCC iron. The introduction of randomly distributed vacancies throughout the lattice results in a dampening of the small $|\mathbf{q}|$ characteristic transverse spin-spin dynamic structure factor peak as well as a shift of these peaks to lower frequency. Splitting of the low $|\mathbf{q}|$ excitation peaks occurs at 5% vacancy concentration, and this behavior is shown to be caused by the formation of vacancy pairs. The constituent components of split transverse excitations are shown to interact with those of differing $|\mathbf{q}|$ through observation of the longitudinal dynamic structure factor.

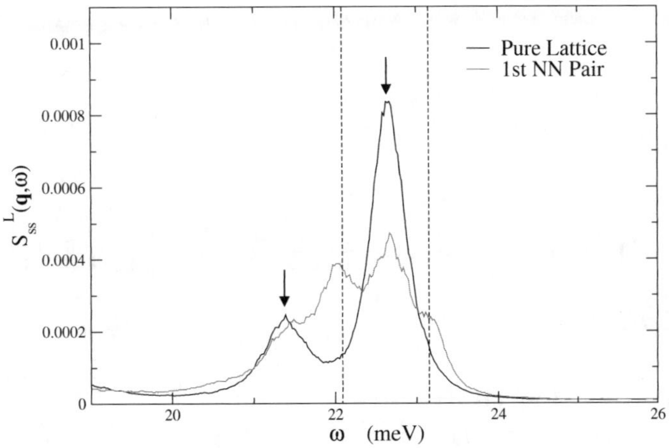


**Figure 3.** Longitudinal spin-spin dynamic structure factor for $|\mathbf{q}| = 0.365\text{Å}^{-1}$ along the $(q, 0, 0)$ direction in an L=8 system. Arrows indicate the predicted locations of the two-spin-wave annihilation peaks for pure lattice simulations. The dashed vertical lines (- - - -) indicate predicted locations of two such peaks in the system containing a nearest neighbor vacancy pair.

## 5. Acknowledgments

This material is based upon work (M.M.) supported by the U. S. Department of Energy, Office of Science, Office of Workforce Development for Teachers and Scientists, Office of Science Graduate Student Research (SCGSR) Program. The SCGSR Program is administered by the Oak Ridge Institute for Science and Education for the DOE under contract name DE-SC0014664. Part of this work (M.E. and G. M. S.) was supported by the Materials Sciences & Engineering Division of the Office of Basic Energy Sciences, U.S. Department of Energy. This study was supported in part by resources and technical expertise from the Georgia Advanced Computing Resource Center, a partnership between the University of Georgia's Office of the Vice President for Research and Office of the Vice President for Information Technology. This research used resources of the Oak Ridge Leadership Computing Facility at ORNL, which is supported by the Office of Science of the U.S. Department of Energy under Contract No. DE-AC05-00OR22725.

## References

[1] Rapaport D C 2004 *The Art of Molecular Dynamics Simulation* 2nd ed (New York: Cambridge University Press)
[2] Allen M P and Tildesley D J 1989 *Computer Simulation of Liquids* (Oxford University Press)
[3] Gerling R W and Landau D P 1990 *Phys. Rev. B* **41** 7139
[4] Hou Z, Landau D P, Stocks G M and Brown G 2015 *Phys. Rev. B* **91** 064417
[5] Tsai S H, Bunker A and Landau D P 2000 *Phys. Rev. B* **61** 333
[6] Yin J, Eisenbach M, Nicholson D M and Rusanu A 2012 *Phys. Rev. B* **86** 214423
[7] Perera D, Eisenbach M, Nicholson D M, Stocks G M and Landau D P 2016 *Phys. Rev. B* **93**(6) 060402
[8] Fransens J R, Keriem M S A E and Pleiter F 1991 *J. Phys.: Condens. Matter* **3** 9871
[9] Dudarev S L and Derlet P M 2005 *J. Phys.: Condens. Matter* **17** 7097
[10] Ma P W, Woo C H and Dudarev S L 2008 *Phys. Rev. B* **78** 024434
[11] Omelyan I P, Mryglod I M and Folk R 2001 *Phys. Rev. Lett.* **86** 898
[12] Lynn J W 1975 *Phys. Rev. B* **11** 2624
[13] Perera D, Nicholson D M, Eisenbach M, Stocks G M and Landau D P 2017 *Phys. Rev. B* **95**(1) 014431
[14] Bunker A and Landau D P 2000 *Phys. Rev. Lett.* **85** 2601

IOP Conf. Series: Journal of Physics: Conf. Series **921** (2017) 012008 doi:10.1088/1742-6596/921/1/012008

# A study of vibrating nanotubes with additional adsorbed masses

**Joan Adler and Omri Adler**

Physics Department, Technion, Haifa, Israel, 32000

E-mail: phr76ja@technion.ac.il

**Abstract.**

We describe calculations of the electronic density surrounding strained nanotubes. These are then used to estimate the nanotube wall width. This width is an essential parameter for the analysis of the nanotube vibrations. By studying the effect of additional adsorbed molecules on the nanotubes' vibrations and their frequency changes we can deduce the molecules' mass. Our calculations show that the strain does not greatly affect the nanotube width, but the vibrations change sufficiently for the mass to be detected.

## 1. Introduction

Vibrating nanotubes can be used as NEMS (Nano-Electro-Mechanical Sensors) to weigh molecules such as DNA. In a series of papers [1, 2, 3, 4] we have explored Molecular Dynamical models of such sensors, and developed protocols for such calculations. One of the challenging aspects of vibrating nanotube systems is the need for an accurate value for $h$, the nanotube wall width, and to study how it varies under local strain in the tube bending. A greatly exaggerated image of a vibrating tube, showing the strained bonds in red is shown at the left of Fig. 1. The width is needed to calculate the mass from the measured frequency, and was very controversial for a long time, concerning the way it plays a rôle in Yakobson's paradox concerning the nanotube's Young's modulus. In [1, 5] a value of $h = 0.06$nm was obtained by two distinct analysis approaches to data from multiple simulations. However until now, direct confirmation was not made.

An intial study of sensitivity to added mass was made in [4] where the simple procedure of replacing the $C_{12}$ molecules with $C_{13}$ ones showed zeptogram sensistivity. However, a direct study of actual additional atoms or molecules was not made until the present project.

In the present project the previous indirect deductions have been confirmed with explicit simulations. No surprises were found, but the road to direct measuring of the molecular mass was rockier than expected. We report below on results of a direct confirmation of the nanotube width value, for strained tubes, and describe the frequency change with adsorbed atoms. In Fig. 1 at right, we show an example of a nanotube with an added half-ring illustrating only the stretched bonds without exaggerated vibrations. In all our vibrational simulations the $y$ axiz is chosen to be along the tube and the $x$ and $z$ directions are perpendicular.

IOP Conf. Series: Journal of Physics: Conf. Series **921** (2017) 012008 doi:10.1088/1742-6596/921/1/012008

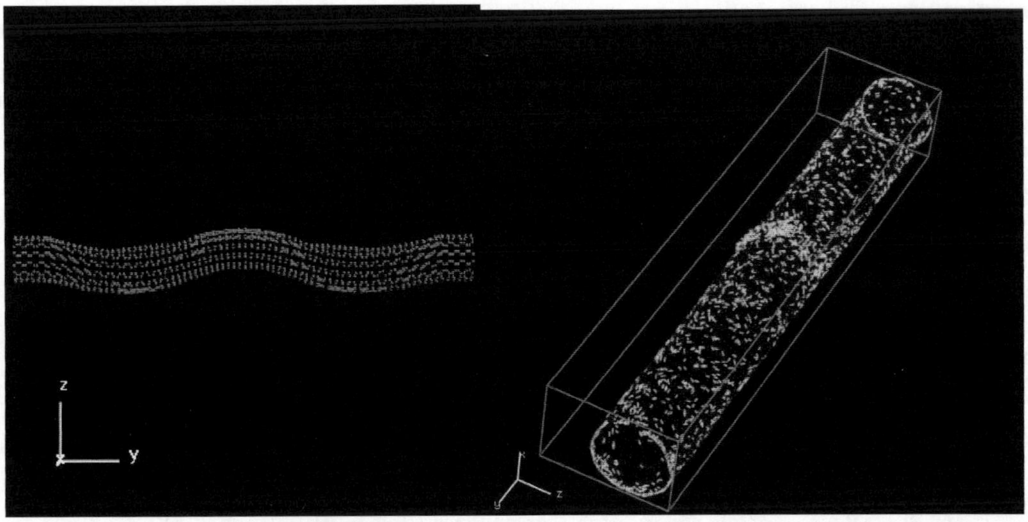


**Figure 1.** Vibrating nanotube with a top-to-bottom color gradient and extended bonds shown in red at left, nanotube with several additional rings and extended bonds in green shown at right. the scale of the vibrational amplitude is much larger than on the right.

## 2. Width measurement

Electronic density is three dimensional, and varies in density as functions of distance from the nucleus and of angle. In two earlier papers, calculation of electronic density surrounding nanotubes was studied within DFT by Quantum Espresso [6, 7, 8]. Its vizualization was carried out in the public domain AViz (Atomistic Vizualization) package [9, 10], developed at the Technion and now part of the FP7 SimPhoNy project [11]. However actual calculations of the width of the envelope surrounding the nanotube were not made systematically for different strains. An image color-coded for different levels of electronic density is shown in Fig. 2 for a single ring and an image with a color coding selected for stereo viewing is shown at right.

We explored how (and if) the width changes for different strains. In Fig. 3 we show the maximun thickness as a function of strain at left and the thickness as a function of density at right. There is very little change in thickness, if anything an increase as strain increases. We hypthesize that this is due to the increasing distance between the positive nuclei (on the $y$ axis) which makes the attractive electrostatic force weaker (on the radial $x$ and $z$ axes) per length unit. In the right frame of the same figure we show that the maximum reaches close to the $h = 0.06$nm that was calculated.

## 3. Change in frequency for additional mass

This section of our study was less straightforward, because after success with substitutional changes in mass in [4] we were unprepared for the event that external adhesion to a vibrating tube would not be so simple. We found that narrow vibrating tubes with a high curvature tend to shake of their burdens. Our experimental collaborators, found similar issues and they suggested decreasing the nanotube's curvature. This led to a delay while revalidating the code and analysis for much thicker tubes. Once this was achieved progress was quite rapid. One wide tube, with an attached half ring of carbon atoms was shown in Fig. 1 above. A full discussion of testing several attached configurations can be found in [13]. Unless an entire ring is added as was the case in [5], where there was clear evidence of a decrease in frequency with an increase in weight, the symmetry between the $x$ and $z$ directions is broken, so we have less statistics for

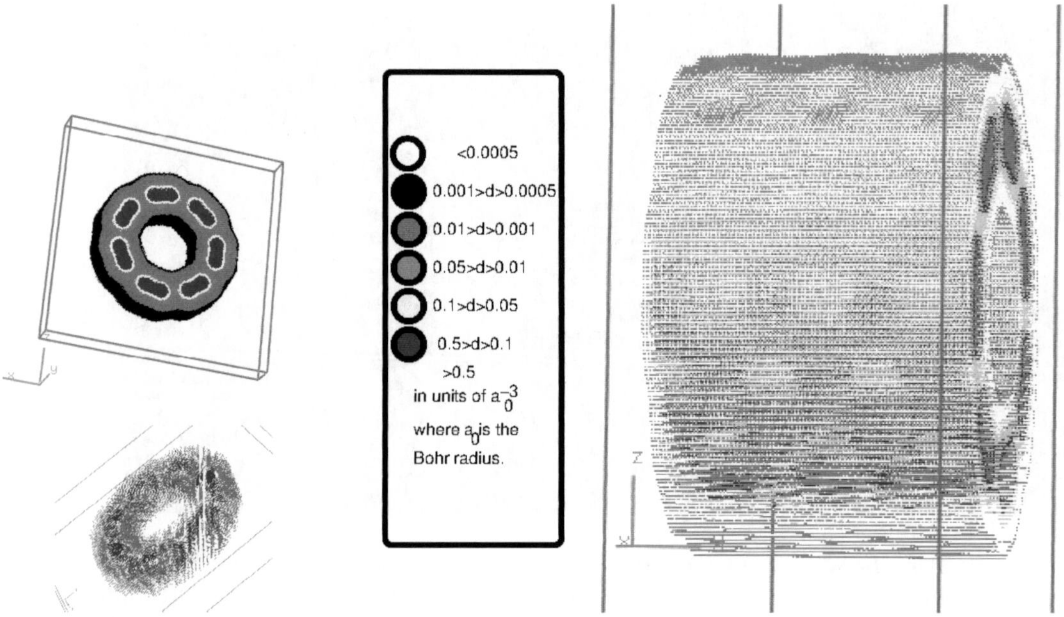


**Figure 2.** Left: Electronic density of a nanotube with a color key, showing a dense image at upper left and a diluted one at lower left; Right: AViz stereo image (with different color code selected for stereo visibility) of density above $0.0005a_0^{-3}$, where $a_0$ is the Bohr radius.

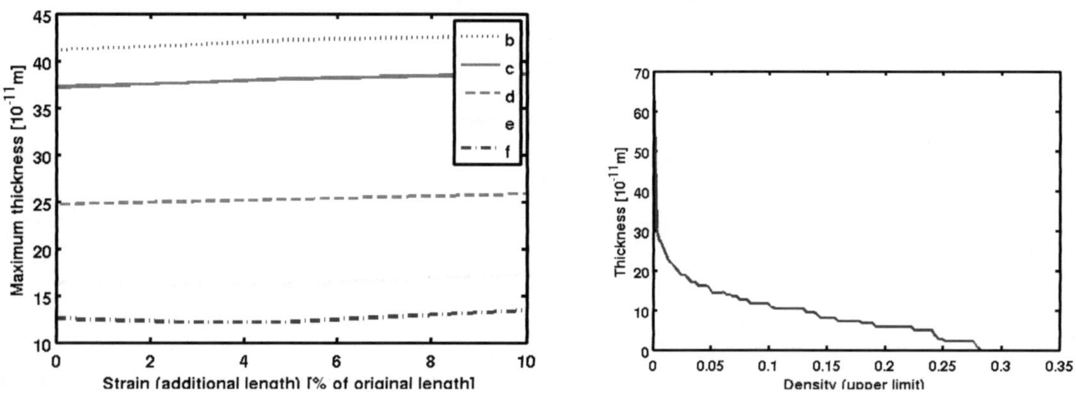


**Figure 3.** Maximum thickness as a function of density at left and thickness as a function of density for a strain of 2.5%

each case, as behavior can differ in the perpendicular directions. There are also considerations of a possible increase in frequency due the contact area between the added rings and the nanotube and a resultant change in elasticity. Numerical estimates for these predictions are given below and details can be found in [13].

One especially interesting case is the study of several different half ring attachments on a (30,30) armchair nanotube with 15060 C atoms. The nanotube was prepared in the same manner as the thinner ones [1], but needed to be longer to vibrate well. The frequency was analysed with a MATLAB code [13] that took its Fourier transform and gave a spectrum of resonance frequencies. The code validation for a long but thin unadorned tube showed that this bare tube

IOP Conf. Series: Journal of Physics: Conf. Series **921** (2017) 012008 doi:10.1088/1742-6596/921/1/012008

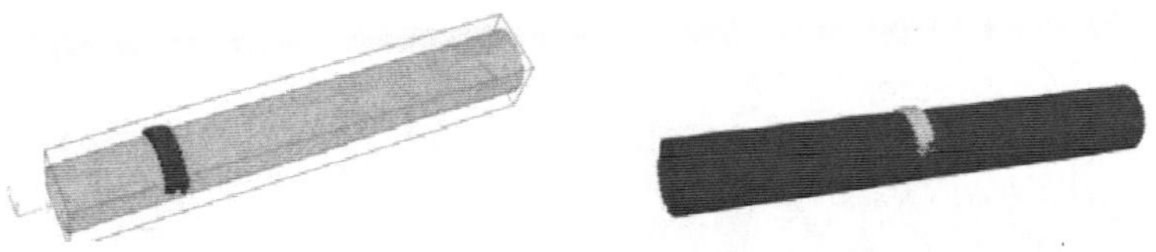

**Figure 4.** Two nanotubes with additional half rings shown in contrasting colors.

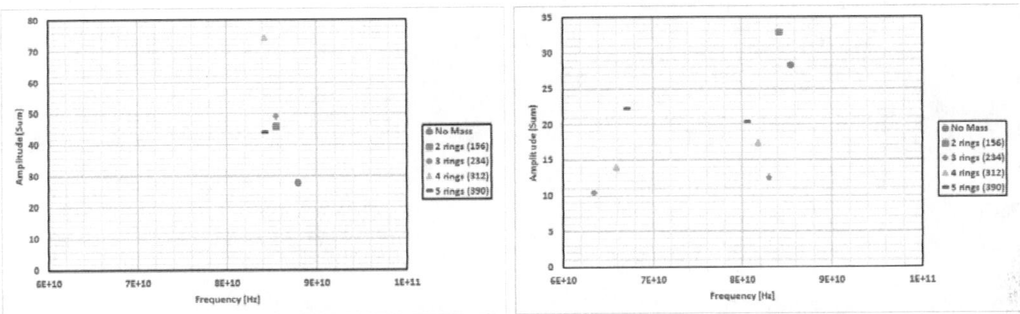


**Figure 5.** Values of the lowest frequencies in the $z$ direction on the left and $x$ direction on the right.

| | No mass | 2 rings | 3 rings | 4 rings | 5 rings |
|---|---|---|---|---|---|
| Number of atoms | 0 | 156 | 234 | 312 | 390 |
| Mass [Kg] | 0.000E+00 | 3.120E-24 | 4.680E-24 | 6.240E-24 | 7.800E-24 |
| Original Theory [Hz] | 8.547E+10 | 8.503E+10 | 8.481E+10 | 8.459E+10 | 8.437E+10 |
| Improved Theory 1 [Hz] | 8.547E+10 | 8.422E+10 | 8.362E+10 | 8.303E+10 | 8.245E+10 |
| Improved Theory 2 [Hz] | 8.684E+10 | 8.421E+10 | 8.298E+10 | 8.181E+10 | 8.068E+10 |
| Simulation (X axis) [Hz] | 8.545E+10 | 8.423E+10 | 8.301E+10 | 8.179E+10 | 8.057E+10 |
| Simulation (Z axis) [Hz] | 8.789E+10 | 8.545E+10 | 8.545E+10 | 8.423E+10 | 8.423E+10 |

**Figure 6.** Table of frequencies for different masses under several versions of theoretical predictions.

had a frequency response that extrapolated nicely from the values found for the thinner tubes in [1, 12].

Attachments of half rings with 78 atoms each were placed at different places along the tube, see two realizations in Fig. 4. From many cases [13] we show graphs of the lowest frequencies in the $x$ and $z$ directions for different numbers of additional half rings in Fig. 5. Despite the differing amplitudes, we observe a clear trend that the frequency decreases as adsorbed mass increases. In Fig. 6 we show tabulated results for the lowest mode for number and mass of atoms, compared with theoretical predictions of several variants of Euler-Bernoulli-Timoshenko theoretical predictions [13] introduced above. $x$ and $z$ directional variations are a slight complication. By comparison with the theory we can deduce the added mass and and thereby "weigh" the additional rings.

## 4. Some asides relating to SimPhoNy

The nanotube vibration study is one of the flag-bearing cases for SimPhoNy, whose raison-d'etre is to streamline simulation modelling for experimental applications. The SimPhoNy concept is

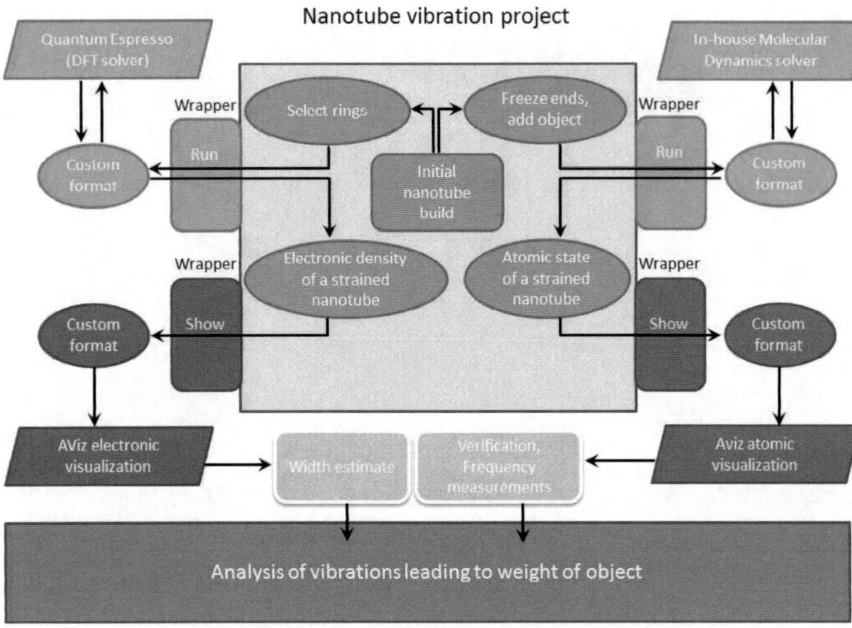


**Figure 7.** SimPhoNy flowchart

that "wrappers" connecting systems on different scales run with different programming and visualization codes which should be "interoperable" and relatively easy for experimental users to run for different parameters. This scheme as applied in the present study is shown in Fig. 7, and a full discussion of this approach in relation to visualization is given in [14].

## 5. Conclusions and prognosis

We have demonstrated that nanotubes with attached masses change frequencies in a systematic manner that enable measurement of the mass as long as one can retain its attachment to the nanotube. Attachment was maintained, by using wider and therefore longer tubes, greatly increasing the computational requirements. It is now time to return to the experimental laboratory to test this result. One proviso we must make is that when this project was begun in 2010, the common Molecular Dynamics code LAMMPS had bugs in the implementation of the Brenner potential that is needed to provide and accurate description of nanotube structure, and so all the calculations were made with an "own code" MD derived from Brenner's original code. The Brenner potential corrections for LAMMPS were recently proposed and so this project will be transformed into a LAMMPS one and then can be integrated into the full SIMPHONY suite, together with the AViz visualizations.

## Acknowledgments

Acknowledgements Supported in part by FP7-NMP-2013-1.4-1 SIMPHONY under grant no. 604005. We thank Yuval Yaish and Sharon Maliniak for discussions on the attachment issues, and Simha Srebnik for the suggestion to draw the attached rings in a different color. We thank Keijo Mattila for assistance in the preparation of the flowchart.

IOP Conf. Series: Journal of Physics: Conf. Series **921** (2017) 012008 doi:10.1088/1742-6596/921/1/012008

## References


[1] Pine P, Yaish Y and Adler J "Simulational and vibrational analysis of thermal oscillations of single-walled carbon nanotubes" 2011 *Phys. Rev.* B **83** 155410

[2] Pine P, Yaish Y and Adler J "Thermal oscillations of structurally distinct single-walled carbon nanotubes", 2011 *Phys. Rev.* B **84** 245409

[3] Pine P, Yaish Y and Adler J "The affect of boundary conditions on the vibrations of single-walled carbon nanotubes" 2011 *J. App. Phys.* **110** 124311

[4] Pine P, Yaish Y and Adler J "Vibrational analysis of thermal oscillations of single-walled carbon nanotubes under axial strain" 2014 *Phys. Rev.* B **89** 115405

[5] Pine P, Yaish Y and Adler J "Simulation of nanosensors based on single walled carbon nanotubes" 2012 *IOP Conference Series* **402** 012002

[6] http://www.quantum-espresso.org and Giannozzi P et al. "QUANTUM ESPRESSO: a modular and open-source software project for quantum simulations of materials" 2009 *J. Phys.* C **21** 395502

[7] Grosso B, Cooper V R, Pine P, Hashibon A, Yaish Y and Adler J 2015 "Visualization of electronic density" *Computer Physics Communications* **195** 1-13

[8] Adler J, Adler O, Kreif M, Cohen O, Grosso B, Hashibon A and Cooper V R 2015 "Mini-review of Electron Density Visualization" *Physics Procedia* **68** 26

[9] http://phony1.technion.ac.il/∼aviz

[10] Adler J 2003 "Visualization in Atomistic and Spin Simulations" *Computers in Science and Engineering"* **5** 61-5.

[11] https://github.com/simphony/AViz

[12] Adler O, Adler J and Yaish Y 2016 "Simulation and vizualization of nanotube vibrations with AViz" poster presented at: CECAM-IRL - Multiscale Simulation: from materials through to Industrial Usage

[13] Adler O 2016 M.Sc. Thesis, Technion

[14] Adler J *et al* to be submitted

# Accuracy and Transferability of *Ab Initio* Electronic Band Structure Calculations for Doped $BiFeO_3$

**Julian Gebhardt**

**Andrew M. Rappe**

Department of Chemistry, University of Pennsylvania, Philadelphia, Pennsylvania 19104, United States

Production Editor, *Journal of Physics: Conference Series*, IOP Publishing, Dirac House, Temple Back, Bristol BS1 6BE, UK

E-mail: `jugeb@sas.upenn.edu`

**Abstract.** $BiFeO_3$ is a multiferroic material and, therefore, highly interesting with respect to future oxide electronics. In order to realize such devices, *pn* junctions need to be fabricated, which are currently impeded by the lack of successful *p*-type doping in this material. In order to guide the numerous research efforts in this field, we recently finished a comprehensive computational study, investigating the influence of many dopants onto the electronic structure of $BiFeO_3$. In order to allow for this large scale *ab initio* study, the computational setup had to be accurate and efficient. Here we discuss the details of this assessment, showing that standard density-functional theory (DFT) yields good structural properties. The obtained electronic structure, however, suffers from well-known shortcomings. By comparing the conventional DFT results for alkali and alkaline-earth metal doping with more accurate hybrid-DFT calculations, we show that, in this case, the problems of standard DFT go beyond a simple systematic error. Conventional DFT shows bad transferability and the more reliable hybrid-DFT has to be chosen for a qualitatively correct prediction of doping induced changes in the electronic structure of $BiFeO_3$.

## 1. Introduction

$BiFeO_3$ (BFO) is a ferroelectric material with a polarization of $P_r \approx 100\,\mu\text{Ccm}^{-2}$[1] below a Curie temperature of $T_C = 1123\,\text{K}$.[2] In addition, the iron centers have a magnetic moment of $\mu_B \approx 3.8\,\mu_\text{B}$[3], which are arranged in a G-typ antiferromagnetic pattern below the Néel temperature of $T_N = 643\,\text{K}$.[2] Therefore, BFO is multiferroic at room temperature and a promising material for oxide electronics. Consequently, BFO has been intensively studied in recent years. Especially doping studies are of importance, as they are the foundation of desired *pn* junctions, which are a vital compound for the envisioned electronics but also photovoltaic devices. While *n*-type doping is readily available and naturally present in most grown BFO crystals due to oxygen vacancies, no successfully *p*-type doped modification has been reported so far. We, therefore, recently carried out a comprehensive computational study, investigating the effects of 31 dopants on the electronic structure of BFO.[4] Here, we assess the computational methodology for doping studies in such a complex material.

IOP Conf. Series: Journal of Physics: Conf. Series **921** (2017) 012009 doi:10.1088/1742-6596/921/1/012009

## 2. Computational Details

A primitive BFO unit cell was considered for initial tests. These tests are carried out as all-electron (AE) calculation employing the FRITZ HABER INSTITUTE AB INITIO MOLECULAR SIMULATIONS package.[5] Total energies and atomic structures are fully relaxed to $10^{-6}$ eV/cell and until forces on ions are below 0.005 eV/Å. The Brillouin zone is sampled by a 3×3×3 Monkhorst-Pack **k**-point grid. [6] We test different density-functional theory (DFT) methods, including two parameterizations of the local-density approximation (LDA),[7, 8] as well as the Perdew–Burke–Ernzerhof (PBE)[9] and the PBEsol[10] generalized-gradient approximations (GGA).

Doping experiments are carried out in a larger (2×2×2) pseudo-cubic unit cell. Substitution of one cation (bismuth or iron) results in a defect concentration of 12.5 %. Structural optimization and initial electronic structure results at the PBE level were carried out using QUANTUM ESPRESSO.[11] In these calculations, core electrons are treated by different flavors of pseudo potential schemes, compiled in the optimized and tested STANDARD SOLID STATE PSEUDOPOTENTIAL database.[12, 13]

Reference electronic structure results are obtained with a hybrid-DFT scheme that uses a fraction of (screened) exact exchange *via* the HSE06 functional[14], with 25 % Hartree-Fock (HF) exchange and a screening parameter $\omega$ of 0.07 $\text{Å}^{-1}$. These calculations were again carried out as AE calculations, using the obtained PBE-optimized geometries.

## 3. Electronic Structure of Pristine $BiFeO_3$

### *3.1. Structural Properties*

BFO can be described in different unit cells. The primitive unit cell of the R3cH structure is rhombohedral ($r$). It is related to a hexagonal ($h$) unit cell with lattice vectors $\mathbf{a}_1^h = a_0^h(\sqrt{3}/2, 1/2, 0)$, $\mathbf{a}_2^h = a_0^h(\sqrt{3}/2, -1/2, 0)$, and $\mathbf{a}_3^h = c_0^h(0, 0, 1)$. Accurate low temperature (5 K) structural data of BFO can be found in the literature,[15] giving lattice vectors for the hexagonal unit cell of $a_0^h = 5.57$ Å and $c_0^h = 13.84$ Å. The deviation of other experimental reports (ICSD:[16] 15299, 75324, 51664, 97591, 157424, 163687, 188396, 191940) is small and errors are within ±0.01 Å and 0.04 Å for $a_0^h$ and $c_0^h$, respectively, with averaged values of 5.58 Å and 13.87 Å. Thus, the low temperature experimental structure of Palewicz *et al.* appears to be of good quality and will serve as reference. The primitive unit cell can be constructed from the lattice parameters of the hexagonal cell with lattice vectors $\mathbf{a}_1^r = (-a_0^h/\sqrt{3}, 0, c_0^h/3)$, $\mathbf{a}_2^r = (\sqrt{3}a_0^h/6, a_0^h/2, c_0^h/3)$, and $\mathbf{a}_3^r = (\sqrt{3}a_0^h/6, -a_0^h/2, c_0^h/3)$.

The experimental structure is compared to various computed structures in table 1. In order to exclude influences of pseudopotentials, we first compare the experimental structure with DFT results using AE calculations. Different flavors of the LDA give similar results, underestimating the lattice constant $a_0^r$ by -2.9 to -3.1 %. PBEsol improves this, with a smaller underestimation of -1.17 %. As it has been observed for many materials, PBE results in an opposite trend, overestimating the lattice constant by 1.12 %. Hence, from the tested standard DFT functionals, PBE performs best, with both PBE and PBEsol yielding structures within reasonable errors from experiment. From this point, two strategies seem to be valid, besides testing for other functionals that are differently or higher parameterized (e.g., meta-GGAs): On the one hand, one could employ one of the functionals that underestimates the lattice constant, supplementing it with a semi-empirical Hubbard correction.[17] The latter usually leads to and expansion of the lattice for oxide materials and also improves the electronic structure by opening the band gap. On the other hand, one may continue with the method that yields the best structural properties at this level of theory, relying on a different scheme for electronic structure properties. We chose the latter (see below), since Hubbard U (and J) corrections can not always be employed in a straight forward way. While it is, in principle, possible to determine a parameter set for a given system from first principles,[18] in practice this does not always provide the best correction to a quantity of interest. In addition, different parameters are often desirable if different quantities are investigated. In fact, for BFO, a variety of different parameters has been applied in the past, with $U_{\text{eff}}$ ranging typically from 0 eV to 4 eV[19–21]. In addition, calculations employing $U = 7$ eV and $J = 0.9$ eV have been proposed.[22] In conjunction with the variety of dopants that should be described equally well, this strategy seems disadvantageous.

**Table 1.** BFO lattice parameter comparing all-electron (AE) and pseudopotential (SSSP) calculations with experimental results.

| Method | $a_0^r$/Å | $a_0^r - a_0^r$(exp.)/% |
|---|---|---|
| Experiment[15] | 5.63 | |
| AE-PBE | 5.69 | 1.12 |
| AE-PBEsol | 5.56 | -1.17 |
| AE-LDA(PZ) | 5.45 | -3.07 |
| AE-LDA(VWN) | 5.46 | -2.89 |
| SSSP-PBE(65 Ry) | 5.67 | 0.72 |
| SSSP-PBE(80 Ry) | 5.73 | 1.80 |
| SSSP-PBE(100 Ry) | 5.70 | 1.36 |
| SSSP-PBE(200 Ry) | 5.70 | 1.35 |

Therefore, we choose the second approach, i.e., we rely on the PBE functional, but only for structural relaxations, whereas electronic structure results are determined in a second step by a different method. For a trade-off between accuracy and computational feasibility, we choose the so called hybrid-DFT, specifically the HSE functional, over more expensive alternatives (see below).

To increase computational efficiency in the first step, the structural optimization at the PBE level, we chose a pseudopotential method employing a plane wave basis set instead of AE calculations. Since our goal is to study a large variety of elements in form of BFO dopants, we choose the highly tested SSSP database.[12] The latter ensures accurate results for the entire periodic table, although with the caveat of being tested for a very costly setup, employing an over-converged plane-wave cutoff of 200 Ry. We tested the respective pseudopotential calculations and its convergence with respect to the plane-wave cutoff (see table 1). It shows that converged structures are obtained for a plane-wave cutoff of 100 Ry. This is, although being reduced by a factor of two compared to the benchmark results of the SSSP,[13] still an unusually extensive basis. In order to ensure accurate results for a wide range of elements, however, we employ this setup in our PBE calculations. Calculations for this setup are in very good agreement with the AE results, indicating the quality of the employed pseudopotentials.

### *3.2. Electronic Structure*

We now turn our discussion to the electronic structure of BFO. The band structure of BFO at the PBE level for our plane wave calculations is shown in figure. 1 (a). It is in very good agreement with the AE result and both band gaps agree within 0.01 eV (see table 2), i.e., the chosen pseudopotentials also do not introduce errors on the electronic structure. It is known, however, that GGA as well as LDA have a band gap problem that leads to the underestimation of computed band gaps.[23] Thus, experimentally observed optical gaps are significantly larger (2.1 eV to 2.8 eV). In order to improve this shortcoming of standard DFT methods, one has to (i) improve the form of the employed density functional or (ii) apply perturbative many-body corrections.[24, 25] We chose the first option, by adding a (fraction) of exact Hartree-Fock exchange to the approximate treatment of exchange and correlation, as implemented in the HSE hybrid-functional. Such a correction is somewhat *ad hoc*, as the fraction of exact exchange is not defined and likely to be system dependent, similar to Hubbard corrections. Therefore, more accurate treatments that go even beyond this hybrid method are desirable. Besides the above mentioned many-body treatment (ii), this is also possible in the spirit of option (i), by employing full exact exchange[26, 27] together with good approximations to the exact correlation contribution.[28] Such methods are, however, computationally very expensive and still under development, hence, not suitable for a broad computational study. Therefore, we have to choose a sensitive parameterization, judging the employed methodology by its performance against reference results. We do so by computing the band structure with the HSE functional. AE calculations at the HSE level give a band gap of 3.17 eV. This is in reasonable agreement with the upper end of the spectrum of experimental optical gaps, overestimating the latter by 0.37 eV. However, we must keep in mind that optical gaps are a lower bond to energy band gaps, which indicates the accuracy of the HSE band gap. This is further validated by comparing to a reported GW band gap of 3.6 eV.[29] The HSE result is in the middle of experimental optical gaps and reference many-body calculations. Despite a quantitatively exact treatment of the gap states in BFO seems to require special care in order to reproduce the observed optical band gaps computationally, the applied hybrid-DFT method seems to provide a band gap description that is qualitatively correct and quantitatively reasonable. We, therefore, choose the hybrid-DFT treatment as reference.

The resulting HSE band structure and density of states (DOS) are shown in figure. 1 (c and d). Besides the changed band gap, significant differences arise in the DOS comparing the HSE to the PBE results. The valence band (VB) region is comprised by hybridized iron (3*d*) and oxygen (2*p*) states that are fully mixed in the PBE case. In contrast, HSE attributes only some iron character to the oxygen dominated VBs, whereas the majority of the iron 3*d* orbitals are located slightly below this set of bands, forming a localized set of *d* states, with only little oxygen contribution. This is in line with previous observations for transition metal *d* states and the reason for the success of Hubbard corrections in these kind of materials. GGA (and LDA) delocalize transition metal *d* states. This is not correct for systems like transition metal oxides such as BFO.[19] Thus, in addition to the commonly observed underestimation of band gaps, standard DFT is additionally problematic whenever a partly filled *d* state is present in the valence region. We will see that this has further important consequences on the results of doped systems, making this level of theory unreliable to even give qualitative trends. This is important, since a mere under- or overestimation of the band gap could be corrected by a simple scissor operation if a more accurate reference (independent of its origin) is known.[30, 31]

**Table 2.** Comparison between optical gaps and computed band gaps for different computational models.

| Method | $E_g$/eV |
|---|---|
| Experiment[32–38] | 2.1 – 2.8 |
| AE-PBE | 1.06 |
| SSSP-PBE | 1.07 |
| AE-HSE | 3.17 |

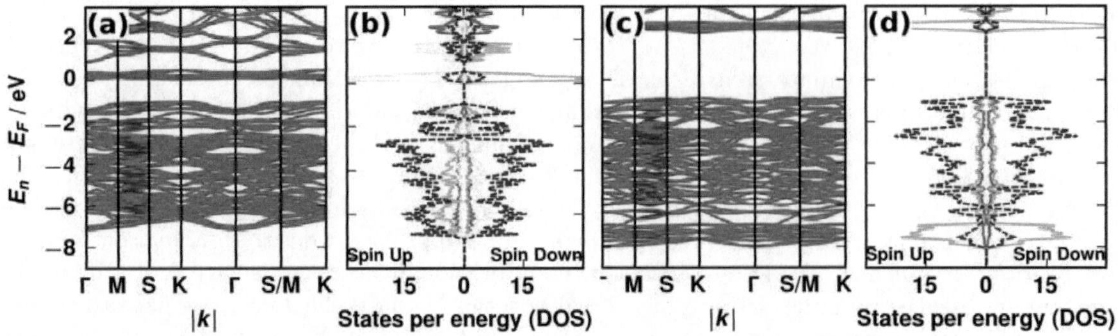


**Figure 1.** Band structure and DOS of undoped BFO at the PBE (a and b) and the HSE (c and d) level, respectively. Bands are colored blue (spin up) and red (spin down) in the band structures (a, c). Contributions to bands are drawn blue (oxygen), orange (iron), and red (bismuth), respectively. In the PBE case, we differentiate between contributions from spin up polarized iron centers (orange) and spin down polarized iron centers (yellow).

## 4. Alkali Metal Doping

With a scheme in hand that is computationally feasible and accurate enough to provide electronic structures quantitatively correct within reasonable error margins and that gives qualitatively correct trends to a high degree of confidence, we can turn to the discussion of doped systems.

We start be investigating the influence of an alkali metal defect on the electronic structure of BFO. The band structure and DOS of sodium doped BFO in the $A$-site is shown in figure. 2. Compared to the pristine case, introducing any alkali metal at $A$- or $B$-sites introduces a $M^+$ defect at a $M^{+3}$ site (here $Na^+$ and $Bi^{+3}$). This means that two electron holes are created. In the band structure we observe this generation of charge carriers by the emergence of two acceptor levels in the band gap. By analyzing the DOS, we find that these acceptor states have mixed iron and oxygen character. In addition with structural changes and the depletion of charge at one iron center, this suggests the localization of both electron holes at one $FeO_6$ octahedron, i.e., the oxidation of $Fe^{+3}$ towards $Fe^{+5}$, as it is discussed in detail in [4]. This process also changes the conduction band (CB) region. In the spin channel opposite to the doping effect, bands are shifted downwards below the CB edge (CBE). In agreement with the oxidation of iron and by analyzing the DOS, it shows that these bands are the unoccupied $d$ states of the reduced iron center, which can be explained by the larger attraction of states by a higher oxidized, i.e., less shielded, atomic core. Figure 2 (a-b) shows the respective band structure and DOS at the PBE level. Hole doping might be concluded by comparing the band dispersion of the VBE and the CBs around 2 eV with the pristine case, identifying bands between the Fermi level and 1 eV as acceptor bands. This however, gives a spurious number of acceptor bands and, comparing the DOS with the HSE results, a physically wrong picture. The lowest acceptor bands are dominated by iron $3d$ states, whereas oxygen contribution is found dominantly in the upper bands. Comparing this with the reference HSE result, we observe several differences. Due to the smaller band gap and the wrong description of $d$ orbitals, the two effects that are introduced because of hole doping are overlapping in the PBE case. The acceptor bands (from the iron center oxidation) are observed in the same spin channel as for HSE. In addition, the down-shift of the iron bands from the CBE in the other spin channel is observed. However, instead of two isolated groups of bands with the acceptor bands located below the shifted CBs, both types of bands are close in energy and the acceptor bands are located closer to the CBE, i.e., the bands appear in the wrong order. This complicates the interpretation of the PBE result, which

IOP Conf. Series: Journal of Physics: Conf. Series **921** (2017) 012009 doi:10.1088/1742-6596/921/1/012009

could lead to miss interpretations. This effect dependents on the size of the chosen alkali metal. For lithium doping, both bands are fully overlapping. For larger alkali metals, the acceptor bands are raised, leading to an increasing separation to the downshifted iron CBs. The spurious stabilization of the unoccupied iron conduction bands can be attributed to the reduced splitting of iron $d$ states by standard DFT, i.e., it is in lined with the occupied $d$ states hybridizing with the oxygen VBs instead of being localized below them. In summary, this means that correct qualitative trends from the PBE results can only be concluded with special care, knowing about the over stabilization of unoccupied iron states.

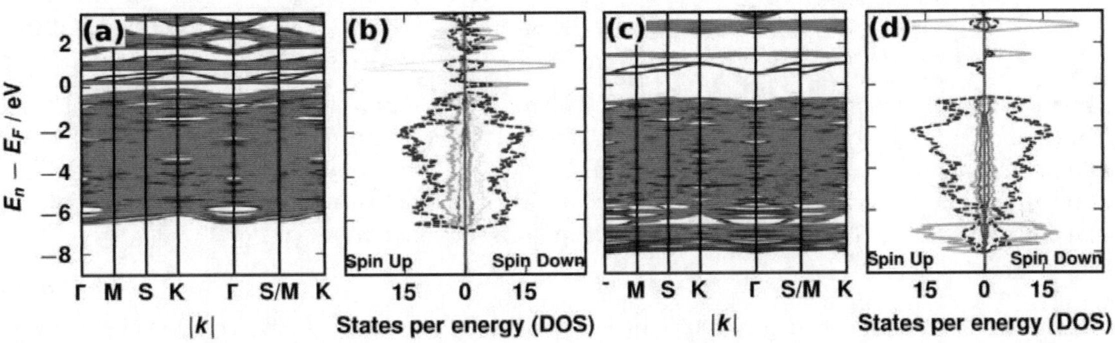


**Figure 2.** Band structure and DOS of BFO with a $Na_A^+$ impurity at the PBE (a and b) and the HSE (c and d) level, respectively. Bands are colored blue (spin up) and red (spin down) in the band structures (a, c). Contributions to bands are drawn blue (oxygen), orange (iron), and red (bismuth), respectively. In the PBE case, we differentiate between contributions from spin up polarized iron centers (orange) and spin down polarized iron centers (yellow).

## 5. Alkaline-earth Metal Doping

Next we discuss the effects of alkaline-earth metal defects. For this we show the band structure and DOS for beryllium in the $B$-site in figure. 3. Similar to the alkali metal dopants, we expect a $M^{+2}$ dopant, resulting in one electron hole generating one acceptor band and, thus, $p$ doping. In fact, we observe great similarities to the effects introduced by $M^{+}$ alkali metal dopants at the HSE level. An acceptor level is observed, containing once again iron and oxygen character. Analyzing the DOS in detail reveals that the electron hole is located at a neighboring iron center, in line with a decreasing octahedron volume and charge depletion. Thus, oxidation of iron to $Fe^{+4}$ is observed. In addition to the doping effect, the reduced charge at the oxidized iron center leads to a stabilization of its empty $d$ orbitals below the CBE, see [4] for details. Looking at the results at the PBE level figure. 3 (a-b), however, acceptor bands are seemingly introduced in both spin channels. Analogously to the alkali metals, this can be attributed to a wrong description of transition metal $d$ states by standard DFT and the too small band gap. The empty iron bands are again stabilized to a degree that they overlap with the actual acceptor band, complicating the interpretation of the observed results.

In figure. 4 we also show the obtained electronic structure of magnesium doped BFO. Here, the application of PBE has even more drastic consequences, resulting in qualitatively wrong predictions from the computed band structure. While the HSE result shows that $Mg^{+2}$ doping affects the band structure analogously to $Be^{+2}$ doping, for PBE we observe partly filled bands at the Fermi level instead. Investigating the DOS shows that the partly filled bands have mixed iron and oxygen character. The iron character is dominantly attributed to spin up polarized iron centers. This suggests a fractional reduction of more than one iron center. This is in line

IOP Conf. Series: Journal of Physics: Conf. Series **921** (2017) 012009 doi:10.1088/1742-6596/921/1/012009

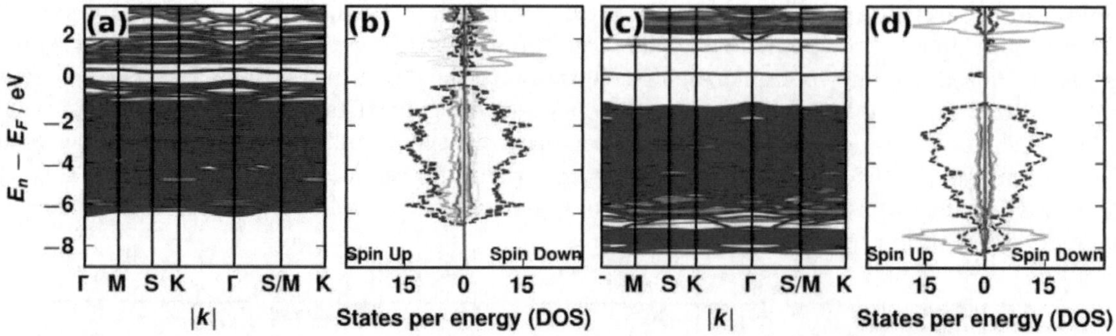


**Figure 3.** Band structure and DOS of BFO with a $Be_B^{+2}$ impurity at the PBE (a and b) and HSE (c and d) level, respectively. Bands are colored blue (spin up) and red (spin down) in the band structures (a, c). Contributions to bands are drawn blue (oxygen), orange (iron), and red (bismuth), respectively. In the PBE case, we differentiate between contributions from spin up polarized iron centers (orange) and spin down polarized iron centers (yellow).

with the tendency of delocalizing charge by standard DFT methods. HSE corrects this behavior, despite using the GGA geometry. This is interesting, as the oxidation of iron centers influences the geometric structure by reducing the oxidized $FeO_6$ octahedra. Partly filled bands at the Fermi level are observed for all alkaline-earth metal dopants besides beryllium at the PBE level, whereas a single acceptor band and a $p$-type doped semiconductor is obtained in each case for HSE. Therefore, standard DFT seems unsuitable to describe the electronic structure of $M^{+2}$ doped BFO. Although structural relaxations with the more accurate hybrid-DFT method seem desirable, qualitatively correct trends are retained using the GGA structures. Therefore, for a large scale study, the chosen two step procedure using the GGA for crystal structures and hybrid-DFT to describe the electronic structure seems required.

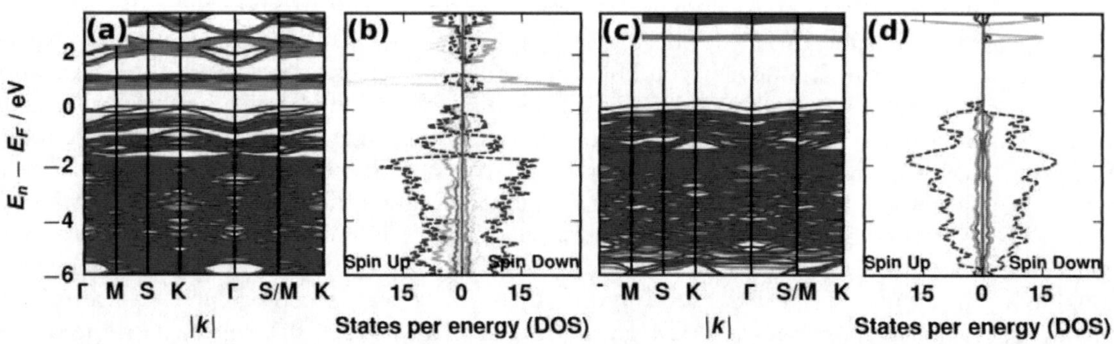


**Figure 4.** Band structure and DOS of BFO with a $Mg_B^{+2}$ impurity at the PBE (a and b) and the HSE (c and d) level, respectively. Bands are colored blue (spin up) and red (spin down) in the band structures (a, c). Contributions to bands are drawn blue (oxygen), orange (iron), and red (bismuth), respectively. In the PBE case, we differentiate between contributions from spin up polarized iron centers (orange) and spin down polarized iron centers (yellow).

## 6. Summary

We assess the accuracy of DFT for the description of BFO. While structural properties seem to be described reasonably well within the GGA, the band gap is underestimated by a large factor. In addition to the well-known band gap underestimation of the GGA, the problematic description of transition metal $d$ shells, which are predicted to be delocalized by standard DFT, intensifies the errors of GGA results. We find that the application of a hybrid-functional provides a corrected electronic structure description and that it provides a good trade-of between computational affordability and sufficient accuracy for a large scale computational study.

We then test our electronic structure treatment by assessing the influences of various impurities, comparing alkali and alkaline-earth metal doping with the undoped case. Doping with metals in oxidation states smaller than three introduces two major effects in the electronic structure: (i) acceptor bands localized on $FeO_6$ octahedra in the vicinity of the dopant and (ii) a downshift of the unoccupied iron $d$ states of the respective iron centers. The ordering of these emerging new bands turns out to be wrong at the PBE level. The problems of electronic structures obtained from the GGA are more pronounced for the case of $M^{+2}$ alkaline-earth metal dopants. Doping with all elements of this group besides beryllium leads to partly filled bands at the GGA level instead of the $p$-doped semiconductors that are predicted by the hybrid-DFT reference method. Thus, qualitatively wrong trends that go beyond a mere underestimation of the band gap are observed using GGA functionals. We, therefore, use a two step procedure, employing GGA calculations only for structural relaxations, whereas the electronic structure is obtained by hybrid-DFT.

## Acknowledgments

J.G. thanks the German Research Foundation for support from Research Fellowship GE 2827/1-1. A.M.R. acknowledges support from the DOE Office of Basic Energy Sciences, under grant number DE-FG02-07ER46431. Computational support is provided by the HPCMO of the U.S. DOD and the NERSC of the U.S. DOE.

## References

[1] G. Catalan, J. F. Scott, *Adv. Mater.* **2009**, *21*, 2463–2485.

[2] Y. E. Roginskaya, Y. Y. Tomashpolskii, Y. N. Venevtsev, V. M. Petrov, G. S. Zhdanov, *Sov. Phys. JETP* **1966**, *23*, 47–51.

[3] I. Sosnowska, W. Schäfer, W. Kockelmann, K. Andersen, I. Troyanchuk, *Appl. Phys. A* **2002**, *74*, S1040–S1042.

[4] J. Gebhardt, A. M. Rappe, **2017**, *Submitted.*

[5] V. Blum, R. Gehrke, F. Hanke, P. Havu, V. Havu, X. Ren, K. Reuter, M. Scheffler, *Comput. Phys. Commun.* **2009**, *180*, 2175–2196.

[6] H. J. Monkhorst, J. D. Pack, *Phys. Rev. B* **1976**, *13*, 5188–5192.

[7] S. H Vosko, L. Wilk, M. Nusair, *Can. J. Phys.* **1980**, *58*, 1200–1211.

[8] J. P. Perdew, A. Zunger, *Phys. Rev. B* **1981**, *23*, 5048–5079.

[9] J. P. Perdew, K Burke, M Ernzerhof, *Phys. Rev. Lett.* **1996**, *77*, 3865–3868.

[10] J. Perdew, A. Ruzsinszky, G. I. Csonka, O. A. Vydrov, G. E. Scuseria, L. A. Constantin, X. Zhou, K. Burke, *Phys. Rev. Lett.* **2008**, *100*, 136406.

[11] P. Giannozzi, S. Baroni, N. Bonini, M. Calandra, R. Car, C. Cavazzoni, D. Ceresoli, G. L. Chiarotti, M. Cococcioni, I. Dabo, A. Dal Corso, S. de Gironcoli, S. Fabris, G. Fratesi, R. Gebauer, U. Gerstmann, C. Gougoussis, A. Kokalj, M. Lazzeri, L. Martin-Samos, N. Marzari, F. Mauri, R. Mazzarello, S. Paolini, A. Pasquarello, L. Paulatto, C. Sbraccia, S. Scandolo, G. Sclauzero, A. P. Seitsonen, A. Smogunov, P. Umari, R. M. Wentzcovitch, *J. Phys. Condens. Matter* **2009**, *21*, 395502.

[12] I. E. Castelli, N. Mounet, A. Marrazzo, G. Prandini, N. Marzari, Standard Solid State Pseudopotentials (SSSP), **2016**.

[13] K. Lejaeghere, G. Bihlmayer, T. Björkman, P. Blaha, S. Blügel, V. Blum, D. Caliste, I. E. Castelli, S. J. Clark, A. D. Corso, S. D. Gironcoli, T. Deutsch, J. K. Dewhurst, I. D. Marco, C. Draxl, M. Du, O. Eriksson, J. A. Flores-Livas, K. F. Garrity, L. Genovese, P. Giannozzi, M. Giantomassi, S. Goedecker, X. Gonze, O. Grånäs, E. K. U. Gross, A. Gulans, F. Gygi, D. R. Hamann, P. J. Hasnip, N. A. W. Holzwarth, D. Iusan, D. B. Jochym, F. Jollet, D. Jones, G. Kresse, K. Koepernik, E. Küçükbenli, Y. O. Kvashnin, I. L. M. Locht, S. Lubeck, M. Marsman, N. Marzari, U. Nitzsche, L. Nordström, T. Ozaki, L. Paulatto, C. J. Pickard, W. Poelmans, M. I. J. Probert, K. Refson, M. Richter, G.-M. Rignanese, S. Saha, M. Scheffler, M. Schlipf, K. Schwarz, S. Sharma, F. Tavazza, P. Thunström, A. Tkatchenko, M. Torrent, D. Vanderbilt, M. J. van Setten, V. V. Speybroeck, J. M. Wills, J. R. Yates, G.-X. Zhang, S. Cottenier, *Science* **2016**, *351*, 1415.

[14] A. V. Krukau, O. A. Vydrov, A. F. Izmaylov, G. E. Scuseria, *J. Chem. Phys.* **2006**, *125*, 224106.

[15] A Palewicz, I Sosnowska, R Przeniosło, A. W. Hewat, *Acta Phys. Pol. A* **2010**, *117*, 296–301.

[16] FIZ-Karlsruhe, Inorganic Crystal Structure Database (ICSD), **2016**.

[17] V. I. Anisimov, O. Gunnarsson, *Phys. Rev. B* **1991**, *43*, 7570–7574.

[18] M. Cococcioni, S. de Gironcoli, *Phys. Rev. B Condens. Matter Mater. Phys.* **2005**, *71*, 18.

[19] J. B. Neaton, C. Ederer, U. V. Waghmare, N. A. Spaldin, K. M. Rabe, *Phys. Rev. B* **2005**, *71*, 014113.

[20] L. Zhu, K. L. Yao, Z. L. Liu, D. H. Zhang, *Phys. Lett. A* **2009**, *373*, 2374–2381.

[21] H. Feng, *J. Magn. Magn. Mater.* **2010**, *322*, 3755–3759.

[22] S. Ju, T. Y. Cai, G. Y. Guo, *J. Chem. Phys.* **2009**, *130*, 214708.

[23] R. O. Jones, O. Gunnarsson, *Rev. Mod. Phys.* **1989**, *61*, 689–746.

[24] L Hedin, *Phys. Rev* **1965**, *139*, A 796 –A 823.

[25] F. Aryasetiawan, O. Gunnarsson, *Reports Prog. Phys.* **1998**, *61*, 237–312.

[26] J. D. Talman, W. F. Shadwick, *Phys. Rev. A* **1976**, *14*, 36–40.

[27] A. Görling, M. Levy, *Phys. Rev. A* **1994**, *50*, 196–204.

[28] J. Erhard, P. Bleiziffer, A. Görling, *Phys. Rev. Lett.* **2016**, *117*, 143002.

[29] V. Shelke, D. Mazumdar, S. Jesse, S. Kalinin, A. Baddorf, A. Gupta, *New J. Phys.* **2012**, *14*, 053040.

[30] Z. H. Levine, D. C. Allan, *Phys. Rev. Lett.* **1989**, *63*, 1719–1722.

[31] X. Gonze, C. Lee, *Phys. Rev. B* **1997**, *55*, 10355–10367.

[32] S. R. Basu, L. W. Martin, Y. H. Chu, M Gajek, R Ramesh, R. C. Rai, X Xu, J. L. Musfeldt, *Appl. Phys. Lett.* **2008**, *92*, 091905.

[33] F. Gao, Y. Yuan, K. F. Wang, X. Y. Chen, F. Chen, J. M. Liu, Z. F. Ren, *Appl. Phys. Lett.* **2006**, *89*, 102506.

[34] A. G. Gavriliuk, V. V. Struzhkin, I. S. Lyubutin, S. G. Ovchinnikov, M. Y. Hu, P. Chow, *Phys. Rev. B* **2008**, *77*, 155112.

[35] U. A. Joshi, J. S. Jang, P. H. Borse, J. S. Lee, *Appl. Phys. Lett.* **2008**, *92*, 242106.

[36] T. Kanai, S.-i. Ohkoshi, K. Hashimoto, *J. Phys. Chem. Solids* **2003**, *64*, 391–397.

[37] A. Kumar, R. C. Rai, N. J. Podraza, S. Denev, M. Ramirez, Y.-H. Chu, L. W. Martin, J. Ihlefeld, T. Heeg, J. Schubert, D. G. Schlom, J. Orenstein, R. Ramesh, R. W. Collins, J. L. Musfeldt, V. Gopalan, *Appl. Phys. Lett.* **2008**, *92*, 121915.

[38] R. Palai, R. S. Katiyar, H. Schmid, P. Tissot, S. J. Clark, J. Robertson, S. A. T. Redfern, G. Catalan, J. F. Scott, *Phys. Rev. B* **2008**, *77*, 014110.

IOP Conf. Series: Journal of Physics: Conf. Series **921** (2017) 012010 doi:10.1088/1742-6596/921/1/012010

# Structure of transition metal clusters: A force-biased Monte Carlo approach

**Dil K. Limbu and Parthapratim Biswas**

Department of Physics and Astronomy, The University of Southern Mississippi, Hattiesburg, Mississippi 39406, USA

E-mail: dil.limbu@usm.edu

E-mail: partha.biswas@usm.edu (Corresponding author)

**Abstract.** We present a force-biased Monte Carlo (FMC) method for structural modeling of transition metal clusters of Fe, Ni, and Cu with 5 to 60 atoms. By employing the Finnis-Sinclair potential for Fe and the Sutton-Chen potential for Ni and Cu, the total energy of the clusters is minimized using a method that utilizes atomic forces in Monte Carlo simulations. The structural configurations of the clusters obtained from this biased Monte Carlo approach are analyzed and compared with the same from the Cambridge Cluster Database (CCD). The results show that the total-energy of the FMC clusters is very close to the corresponding value of the CCD clusters as listed in the Cambridge Cluster Database. A comparison of the FMC and CCD clusters is presented by computing the pair-correlation function, the bond-angle distribution, and the distribution of atomic-coordination numbers in the first-coordination shell, which provide information about the two-body and three-body correlation functions, the local atomic structure, and the bonding environment of the atoms in the clusters.

## 1. Introduction

In recent years there has been a rapid development in the global optimization techniques, which include state-of-the-art evolutionary search strategies to the population-based swarm intelligence and differential-evolution approaches [1]. In spite of this development, the Monte Carlo (MC) methods–based on the Metropolis or related algorithms–continue to play a major role in addressing difficult optimization problems in science and engineering. Since the calculation of local gradients of a potential is computationally more complex than the evaluation of the total energy of a system, the MC methods are often preferred in many optimization problems where local gradients are either not available or computationally too expensive to compute. However, the advantage of the MC methods is often outweighed by their slow convergence behavior, which require a longer simulation time to produce results with the desired accuracy when compared to the Newton-like and Conjugate-Direction methods. The main purpose of the paper is to explore the usefulness of employing atomic forces in Monte Carlo simulations. In particular, we study here the application of a simple gradient-based Monte Carlo method, originally introduced by Rossky *et al.* [2, 3], to optimize the transition metal clusters of Fe, Ni, and Cu and compare the results with the putative global minimum of these clusters reported in recent literature [4, 5].

Transition metal clusters (TMCs) have been studied extensively from computational and experimental points of view [4–8]. Theoretical efforts to study TMCs include empirical Monte Carlo and molecular-dynamics simulations to semi-empirical tight-binding and full *ab initio* calculations. Among numerous studies, the structure of Fe clusters obtained from the Finnis-Sinclair potential [9, 10] and that of Ni and Cu clusters bound by the Sutton-Chen potential [11] are of particular interest. The putative global minima of Ni and Cu clusters of varying sizes have been studied systematically by Doye and Wales [5] using the Sutton-Chen potential. Likewise, Elliott *et al.* [4] have recently addressed the computation of the global minima of several Fe clusters using the Finnis-Sinclair potential. In both the cases, the authors have employed an improved version of the basin-hopping algorithm of Li and Scheraga [12] using Monte Carlo simulations coupled with the Conjugate-Gradient optimization [13]. The total energy and structures of these clusters are available from the Cambridge Cluster Database (CCD). In the following, we refer these clusters as the CCD clusters and use them as a benchmark for the comparison of the total energy and structure of Fe, Ni, and Cu clusters obtained from the force-biased Monte Carlo (FMC) simulations presented here.

## 2. Computational Method

We began by generating a random configuration consisting of N atoms and computing the total energy of the configuration using an appropriate interatomic potential. Here, we chose to employ the Finnis-Sinclair potential [9] for Fe clusters and the Sutton-Chen potential [11] for Ni and Cu clusters. The initial configuration was equilibrated at temperature $T$=3000 K for $10^5$ Monte Carlo steps (MCS). Thereafter, the temperature was decreased by a factor of 0.99 and the system was equilibrated for $10^5$ MCS at each temperature until the final temperature of the system was reduced to 1 K. The total-energy relaxation was achieved in two steps: a) first, we computed the total force on each atom in the initial state $n$ and displaced a randomly selected atom at site $i$ from an initial state $n$ to a proposed state $m$ by [2, 14],

$$\mathbf{\Delta r_i^{mn}} = \alpha\, \delta \mathbf{r_i^{mn}} + \beta\, A\, \mathbf{f_i^n}. \tag{1}$$

Here, $\beta = \frac{1}{k_B T}$ and, $\alpha$ and $A$ are parameters that determine the length of the random displacement ($\alpha\, \delta \mathbf{r_i^{mn}}$) and the contribution from the potential gradient (-$\mathbf{f_i^n}$) at site $i$ in generating a proposed configuration $m$, respectively. The displacement $\delta \mathbf{r_i^{mn}}$ is generally, but not necessarily, drawn from a Gaussian distribution with a zero mean and a variance $2A$; b) second, the proposed configuration obtained in step (a) was either accepted or rejected. Following Rossky *et al.* [2], and Allen and Tildesley [3], one can show that the proposed move in Eq. (1) can be accepted with the probability $P_{mn} = min[1, \exp(-\beta\, \Delta E_i^{mn})]$, where

$$\Delta E_i^{mn} = \delta E^{mn} + \left[\frac{1}{2}(\mathbf{f_i^n} + \mathbf{f_i^m}) \cdot \mathbf{\Delta r_i^{mn}} + \frac{\beta A}{4}\{(\delta \mathbf{f_i^{mn}})^2 + 2\mathbf{f_i^n} \cdot \delta \mathbf{f_i^{mn}}\}\right] \tag{2}$$

$$\delta E^{mn} = E^m - E^n, \qquad \delta \mathbf{f_i^{mn}} = \mathbf{f_i^m} - \mathbf{f_i^n}.$$

In Eq. (2) above, $E^n$ and $E^m$ are the total energy of the system in the initial state and the proposed state, respectively. Likewise, $\mathbf{f_i^n}$ and $\mathbf{f_i^m}$ are the total force on an atom at site $i$ before and after the displacement, respectively. In this work, we moved one atom at a time but it is straightforward to move a group of atoms simultaneously by ensuring that the change of total energy, $\Delta E^{mn}$, associated with multi-atom moves, is properly evaluated. To improve the acceptance rate, we adjusted the step length by restricting $\alpha$ between 0.001 Å at 1 K and 0.05 Å at 3000 K. The value of $A$ was chosen in such a way that $\beta A \approx$ 4–5 $\times 10^{-3}$ and $\delta \mathbf{r_i^{mn}}$ was a random number drawn from a uniform distribution between -1 and +1. In this preliminary study, we made no attempts to optimize the values of $\alpha$ and $A$, which can be adjusted during simulations to improve the efficiency of the method.

## 3. Results and Discussions

### *3.1. Total energy of Fe, Ni, and Cu clusters*

In this section, we present the results from our simulations with an emphasis on the structural properties of the clusters. Following a comparison of the total-energy values of the FMC clusters with the same from the Cambridge Cluster Database [4, 5], we analyze the two- and three-body correlation functions

**Table 1.** Total energy of Fe, Ni, and Cu clusters from the FMC and CCD simulations.

| $N$ | Fe (eV) FMC (CCD) | Ni (eV) FMC (CCD) | Cu (eV) FMC (CCD) |
|---|---|---|---|
| 5 | -11.8598 (-11.8598) | -14.6033 (-14.6033) | -11.5120 (-11.5120) |
| 10 | -28.5357 (-28.5357) | -32.5487 (-32.5487) | -25.6584 (-25.6585) |
| 15 | -46.6371 (-46.6375) | -51.3230 (-51.3231) | -40.4584 (-40.4586) |
| 20 | -64.8373 (-64.8385) | -69.9806 (-69.9833) | -55.1663 (-55.1686) |
| 25 | -82.9394 (-82.9402) | -89.0701 (-89.0708) | -70.2151 (-70.2155) |
| 30 | -101.4469 (-101.4513) | -108.4284 (-108.4296) | -85.4753 (-85.4762) |
| 35 | -119.5937 (-119.5971) | -127.9645 (-127.9657) | -100.8758 (-100.8767) |
| 40 | -138.5100 (-138.5112) | -147.5979 (-147.6000) | -116.3534 (-116.3547) |
| 45 | -156.6982 (-156.6997) | -167.1068 (-167.1649) | -131.7262 (-131.7780) |
| 50 | -175.4342 (-175.4720) | -187.0454 (-187.2391) | -147.6007 (-147.6026) |
| 55 | -194.6847 (-194.6868) | -207.6107 (-207.6135) | -163.6617 (-163.6640) |
| 60 | -214.4256 (-214.4278) | -226.8486 (-226.9008) | -178.8274 (-178.8684) |

and show that the three-dimensional distribution of atoms in the FMC and CCD clusters are very close to each other as far as these correlation functions are concerned. We then address the distribution of the coordination numbers of the atoms in order to obtain further information about the atomic arrangement in the first-coordination shell of atoms. Table 1 lists the total energy of Fe, Ni, and Cu clusters from the FMC simulations. The corresponding total energy of the CCD clusters is also indicated in Table 1. A comparison of the total-energy values suggests that the maximum absolute deviation of the total energy (between the FMC and CCD structures) is less than 0.2 eV, which translates a percentage deviation of 0.1% (see $Ni_{50}$). Figure 1 presents the total-energy difference, $\Delta E = E_{\rm FMC} - E_{\rm CCD}$, between the FMC and CCD clusters. For Fe clusters of up to 45 atoms, the FMC and CCD structures have practically the same energy, whereas the energy of the remaining few clusters with $N > 45$ shows a minor deviation of up to 0.05 eV from the CCD configuration. A similar observation applies for Cu clusters in Fig. 1(c) and Ni clusters in Fig. 1(b) with the exception of $Ni_{50}$, as mentioned earlier. It is evident from Table 1 that the optimized FMC clusters are as stable as the corresponding CCD clusters as far as the total energy of the clusters is concerned.

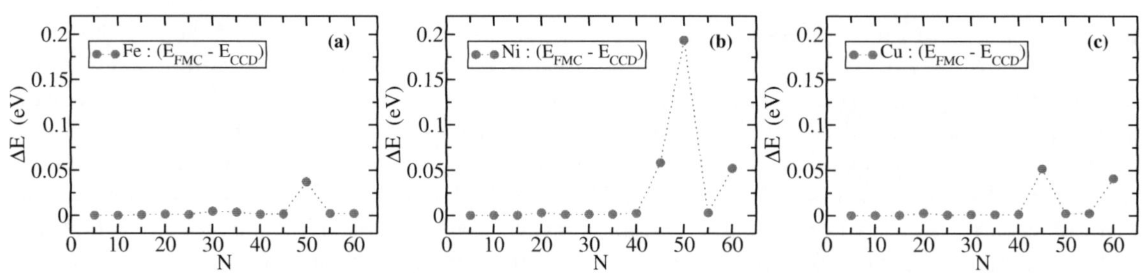


**Figure 1.** Total-energy difference ($\Delta E$) as a function of the cluster size ($N$) for a) Fe, b) Ni, and c) Cu clusters.

### 3.2. *Local structure and atomic environment*

We now discuss the two- and three-body correlation functions in order to examine the three-dimensional distribution of atoms in the clusters. Toward that end, the pair-correlation functions for Fe, Ni, and

Cu clusters are plotted in Fig. 2 along with the corresponding pair-correlation functions for the CCD clusters. The bond-angle distributions for these clusters are also plotted in Fig. 3 for comparison. The results indicate that the two-body and three-body correlation functions from the FMC structures match very closely with their CCD counterparts reflecting structural similarities from the point of view of radial and bond-angle distribution functions.

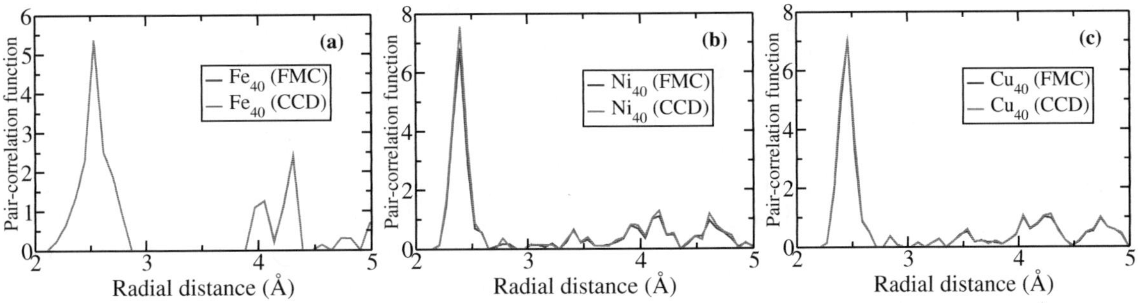


**Figure 2.** The pair-correlation functions for a) Fe, b) Ni, and c) Cu clusters from the FMC simulations. The corresponding pair-correlation functions for the CCD clusters are also shown for comparison.

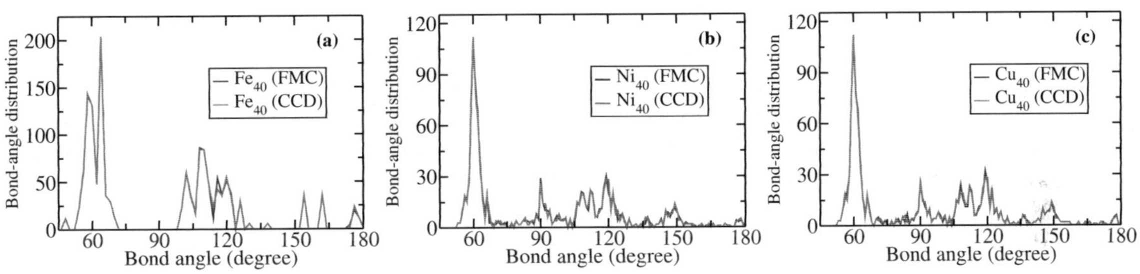


**Figure 3.** Bond-angle distributions for a) Fe, b) Ni, and c) Cu clusters from the FMC and CCD structures.

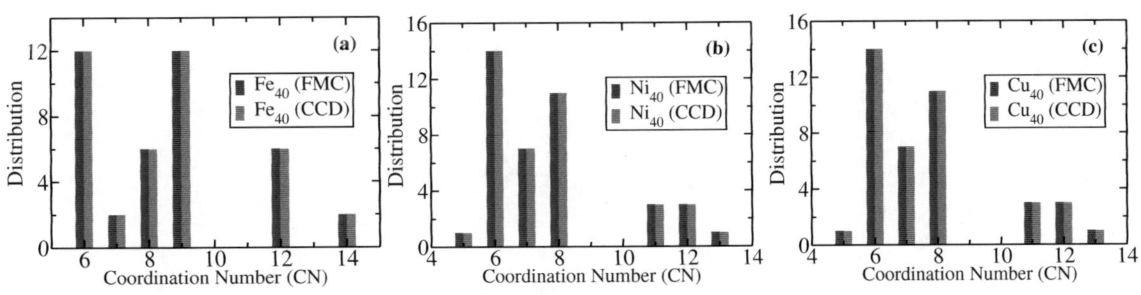


**Figure 4.** The atomic-coordination numbers in a representative 40-atom a) Fe, b) Ni, and c) Cu cluster.

While the pair-correlation function and the bond-angle distribution of a cluster provide considerable information about the local atomic structure and the bonding geometry of the atoms, these distributions cannot uniquely characterize a three-dimensional distribution of atoms in real space. Reverse Monte Carlo simulations of amorphous materials [15–17] – and the recent work on amorphous silicon and amorphous silica [18–21], using hybrid simulation strategies, which employed both the total energy and the atomic forces in structural determination of complex non-crystalline materials – indicate that one needs to include structural information beyond three-body atomic correlations in order to generate a unique atomic arrangement from a given set of structural data from experiments. For this purpose, we have computed the distribution of the atomic-coordination numbers in the first-coordination shell of the atoms and their bonding environment in order to obtain additional information on the atomic arrangement of the clusters beyond the two-body and the reduced three-body correlation functions. In Fig. 4, we have shown the distributions of the atomic-coordination numbers for Fe, Ni, and Cu clusters. The results indicate that not only the average-coordination number of an FMC cluster and a CCD cluster matches with each other, which is evident from the area under the first peak of the pair-correlation function in Fig. 2, but also the full distribution of the atomic-coordination numbers as shown in Fig. 4.

## 4. Conclusion

In this paper, we have presented a Monte Carlo study of total-energy optimization and structural properties of transition metal clusters of Fe, Ni, and Cu. Using a force-biased Monte Carlo method, we have optimized the total energy of the clusters by employing the Finnis-Sinclair potential for Fe and the Sutton-Chen potentials for Ni and Cu. Unlike conventional Monte Carlo simulations, where the acceptance probability of a random displacement of an atom is governed by the energy difference between the final and initial states only, the present method employs both the total-energy difference and atomic forces in determining the acceptance probability associated with the Monte Carlo moves. The total energy of the clusters obtained from the FMC simulations is compared with the data listed in the Cambridge Cluster Database (CCD). The results suggest that the FMC method can produce structural configurations, which are essentially identical to that of the CCD configurations as far as the total energy, the pair-correlation function, the bond-angle distribution, and the atomic-coordination numbers are concerned. The method can be applied to model bulk amorphous solids where one is primarily interested in obtaining a set of atomic configurations that correspond to a few low-lying local minima on the potential energy surface. In a future communication, we shall address this problem using an optimized version of the FMC algorithm.

## Acknowledgments

The work is partially supported by the National Science Foundation (NSF) under Grant No. 1507166.

## References

[1] Engelbrecht A P 2007 *Computational Intelligence – An introduction* (John Wiley and Sons, Ltd)
[2] Rossky P J, Doll J D and Friedman H L 1978 *J. Chem. Phys.* **69** 4628–4633
[3] Allen M P and Tildesley D J 1987 *Computer Simulation of Liquids* (Oxford University Press)
[4] Elliott J, Shibuta Y and Wales D 2009 *Philos. Mag.* **89** 3311–3332
[5] Doye J P K and Wales D J 1998 *New J. Chem.* **22**(7) 733–744
[6] Luo C 2002 *New J. Phys.* **4** 10
[7] Montejano-Carrizales J M, Iñiguez M P, Alonso J A and López M J 1996 *Phys. Rev. B* **54**(8) 5961–5969
[8] Parks E K, Zhu L, Ho J and Riley S J 1995 *J. Chem. Phys.* **102** 7377–7389
[9] Finnis M W and Sinclair J E 1984 *Philos. Mag. A* **50** 45–55
[10] Finnis M W and Sinclair J E 1986 *Philos. Mag. A* **53** 161
[11] Sutton A P and Chen J 1990 *Philos. Mag. Lett.* **61** 139–146
[12] Li Z and Scheraga H A 1987 *Proc. Natl. Acad. Sci. USA* **84** 6611–6615
[13] Wales D J and Doye J P K 1997 *J. Phys. Chem. A* **101** 5111–5116
[14] Pangali C, Rao M and Berne B 1978 *Chem. Phys. Lett.* **55** 413 – 417
[15] Biswas P, Atta-Fynn R and Drabold D A 2004 *Phys. Rev. B* **69**(19) 195207
[16] McGreevy R L 2001 *J. Phys. Condens. Matter* **13** R877
[17] Biswas P, Barkema G T, Mousseau N and vander Weg W F 2001 *Europhys. Lett.* **56** 427–433
[18] Biswas P, Tafen D N and Drabold D A 2005 *Phys. Rev. B* **71** 54204
[19] Pandey A, Biswas P and Drabold D A 2015 *Phys. Rev. B* **92**(15) 155205
[20] Pandey A, Biswas P and Drabold D A 2016 *Sci. Rep.* **6** 33731
[21] Pandey A, Biswas P, Bhattarai B and Drabold D A 2016 *Phys. Rev. B* **94** 235208

# An iterative aggregation and disaggregation approach to the calculation of steady state distributions of continuous processes

**Katja Biswas**

Department of Physics, University of North Georgia, US

E-mail: Katja.Biswas@ung.edu

**Abstract.** A mapping of the process on a continuous configuration space to the symbolic representation of the motion on a discrete state space will be combined with an iterative aggregation and disaggregation (IAD) procedure to obtain steady state distributions of the process. The IAD speeds up the convergence to the unit eigenvector, which is the steady state distribution, by forming smaller aggregated matrices whose unit eigenvector solutions are used to refine approximations of the steady state vector until convergence is reached. This method works very efficiently and can be used together with distributed or parallel computing methods to obtain high resolution images of the steady state distribution of complex atomistic or energy landscape type problems. The method is illustrated in two numerical examples. In the first example the transition matrix is assumed to be known. The second example represents an overdamped Brownian motion process subject to a dichotomously changing external potential.

## 1. Introduction

In recent years the projection of continuous processes onto discrete state space models has gained popularity. This is motivated by the fact that for systems consisting of a large number of atoms (such as bio-polymers) the dynamical process is often too complicated to be understood intuitively from single long time trajectories and in many cases also computationally uneconomical. The latter particularly holds true for processes that evolve on disparate time and length scales. One approach to overcome these problems consists of building coarse discrete models representing the dynamics of the system. They consist of a set of abstract states and are represented by a stochastic matrix whose entries represent the transition between the states. If chosen adequately, this matrix can lead to an intuitive interpretation of important features of the original process. In this context Markov state models (MSM) [1, 2] have gained widespread importance. Constructed in a two-step procedure[1], the original dynamics is first projected onto a large set of microstates from which, following certain criteria, macrostates are build. For realistic bio-polymers, microstate models typically can consist of $10,000$ to $100,000$ states making them hard to tackle computationally.

Here I will use the projection of a continuous process onto a discrete state space together with an iterative aggregation and disaggregation (IAD) procedure to obtain a high-resolution steady state distribution of the process. The iterative aggregation and disaggregation methods are a class of methods [3, 4, 5, 6] known for being able to deal very efficiently with large and sparse stochastic matrices [7, 8]. Iteratively, the IAD methods calculate the eigenvector pertaining to the unit eigenvalue, which is the steady state distribution. Using an approximation of the steady state vector the stochastic matrix is aggregated to form a smaller matrix. The eigenvector problem is then solved on the smaller state space and, via a disaggregation step, its solution is used to refine the approximation of the steady state vector. In the projection of processes from the continuous phase- or configuration- space onto a discrete state space representation consisting of a large number of states, one will find that in majority of the cases a state will be connected to only a very limited number of other states. This leads to a large and sparse stochastic matrix [6], for which the unit eigenvector can be effectively computed using the IAD methods. Therefore, combining the IAD with a phase- or configuration- space mapping procedure may provide a valuable way to deal with problems on the microstate level and may give useful information for macrostate clustering.

The article is structured as follows. In section 2 the method will be introduced, starting with the mapping procedure from the continuous process onto the discrete state space, followed by the IAD method. In section 3, the method will be illustrated at two numerical examples. 3.1 shows a simple case for which a $(5 \times 5)$ transition matrix is known. This didactical example will be used to illustrate the steps in the iterative calculation of the steady state distribution. In section 3.2, the method will be shown at the Brownian motion process of a particle subject to a one dimensional potential which changes its shape with a dichotomous correlation in time. In this example the transition matrix P will be obtained using distributed computing and compared with those obtained from a full simulation.

## 2. Method

Dynamical processes which are described by continuous phase- or configuration- space trajectories can be represented by the probabilities of transition between an abstract set of states. For this purpose, the accessible phase- or configuration- space needs to be projected onto a set of discrete states which I will label $i$, here $i = 1, \ldots, m$ and $m$ is the total number of states. This can be done by dividing the phase- or configuration- space into $m$ non-overlapping intervals labeled $\zeta_i$ [see fig. (1)]. A trajectory $\Gamma_l$ is then said to be found in state $i$ at time $t$ if $h_{\zeta_i}[\Gamma_l(t)] = 1$, where

$$h_{\zeta_i}[\Gamma_l(t)] = \begin{cases} 1 & \text{, if } \Gamma_l(t) \in \zeta_i \\ 0 & \text{, otherwise} \end{cases} \tag{1}$$

is the indicator function. This is used to measure the conditional probabilities $p(j|i)$ representing the probabilities at which a system that resides in state $i$ at time $t$ will have made a transition from state $i$ to state $j$ within a single time step $\mathrm{d}t$. They are given by[1]

$$p(j|i) = \begin{cases} \frac{\langle h_{\zeta_j}[\Gamma_l(t+\mathrm{d}t)] h_{\zeta_i}[\Gamma_l(t)]\rangle}{\langle h_{\zeta_i}[\Gamma_l(t)]\rangle} & \text{, if } \langle h_{\zeta_i}[\Gamma_l(t)]\rangle \neq 0 \\ 0 & \text{, otherwise} \end{cases} , \tag{2}$$

where the angular brackets $\langle \ldots \rangle$ denote the time and ensemble average. They can be used to set up the transition matrix $\mathbf{P}$ whose elements, the transition probabilities $p_{ij}$, are defined as

$$p_{ij} = \begin{cases} p(j|i) & \text{, if } j \neq i \\ 1 - \sum_{j \neq i} p(j|i) & \text{, if } j = i \end{cases} . \tag{3}$$

The $p_{ij}$, with $j \neq i$, denote the probabilities of the system making a transition from state $i$ to $j$ within a single time step $\mathrm{d}t$. The elements $p_{ii}$ denote the probability of a system that resides in

[1] Eqn. (2) is an expression of $p(j, t+\mathrm{d}t|i, t)p(i, t) = p(j, t+\mathrm{d}t; i, t)$, where $p(j, t+\mathrm{d}t; i, t)$ is the joined probability of finding the system at time $t$ in state $i$ and at time $t + dt$ in state $j$. More details concerning the notation of the mapping procedure can be found in [9, 10].

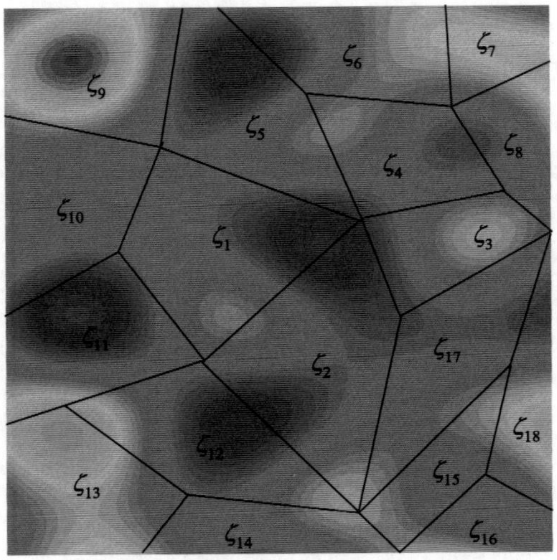


**Figure 1.** Illustration of intervals on an arbitrary 2-dim energylandscape.

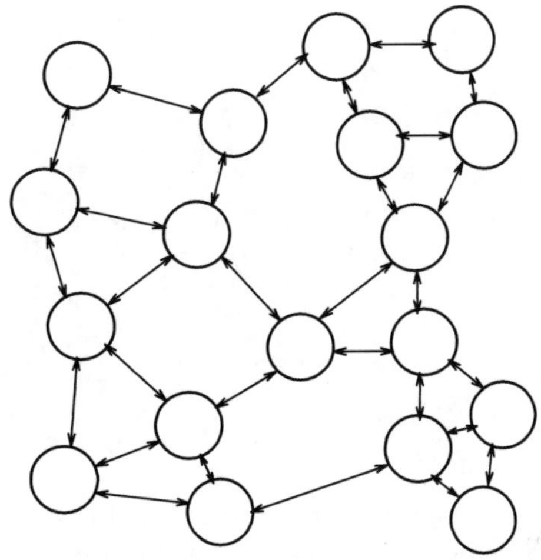

**Figure 2.** A discrete representation of the dynamics of a system.

state $\zeta_i$ to stay in the same state during the time step d$t$. The transition matrix is then given as

$$\mathbf{P} = \begin{pmatrix} p_{11} & p_{12} & p_{13} & p_{14} & \cdots & p_{1m} \\ p_{21} & p_{22} & p_{23} & p_{24} & \cdots & p_{2m} \\ p_{31} & p_{32} & p_{33} & p_{34} & \cdots & p_{3m} \\ p_{41} & p_{42} & p_{43} & p_{44} & \cdots & p_{4m} \\ \vdots & \vdots & \vdots & \vdots & \ddots & \vdots \\ p_{m1} & p_{m2} & p_{m3} & p_{m4} & \cdots & p_{mm} \end{pmatrix}. \tag{4}$$

For a set of $m$ discrete states $i$ the matrix $\mathbf{P}$ has dimension $(m \times m)$. See fig. (2) for a visualization. Particularly for partitions of the original continuous phase- or state-space which lead to a very large number of states, finding the steady state distribution can be a very arduous task. The steady state distribution is given by the eigenvector corresponding to the unit eigenvalue and corresponds to a situation in which the probabilities of occupation of the individual states are not changing with respect to time. In the following the steady state distribution will be obtained using an iterative aggregation and disaggregation procedure.
In order to apply the procedure, the matrix $\mathbf{P}$ has to be partitioned into block matrices. Roughly, this can be interpreted as outlining macrostates, but with the difference that the information about the transitions between the smaller states is preserved at this step [compare to fig. (3)], i.e.

$$\mathbf{P} = \begin{pmatrix} \begin{bmatrix} p_{11} & p_{12} & p_{13} \\ p_{21} & p_{22} & p_{23} \\ p_{31} & p_{32} & p_{33} \end{bmatrix} & \begin{bmatrix} p_{14} & \cdots \\ p_{24} & \cdots \\ p_{34} & \cdots \end{bmatrix} & \cdots & \begin{bmatrix} p_{1(m-1)} & p_{1m} \\ p_{2(m-1)} & p_{2m} \\ p_{3(m-1)} & p_{3m} \end{bmatrix} \\ \vdots & \vdots & \ddots & \vdots \\ \begin{bmatrix} \vdots & \vdots & \vdots \\ p_{m1} & p_{m2} & p_{m3} \end{bmatrix} & \begin{bmatrix} \vdots & \vdots \\ p_{m4} & \cdots \end{bmatrix} & \cdots & \begin{bmatrix} \vdots & \vdots \\ p_{m(m-1)} & p_{mm} \end{bmatrix} \end{pmatrix} = \begin{pmatrix} \mathbf{P}_{11} & \mathbf{P}_{12} & \cdots & \mathbf{P}_{1\ell} \\ \mathbf{P}_{21} & \mathbf{P}_{22} & \cdots & \mathbf{P}_{2\ell} \\ \vdots & \vdots & \ddots & \vdots \\ \mathbf{P}_{\ell 1} & \mathbf{P}_{\ell 2} & \cdots & \mathbf{P}_{\ell\ell} \end{pmatrix}.$$

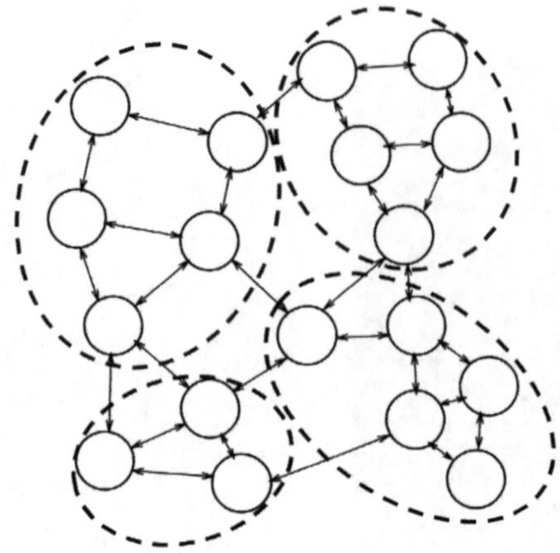

**Figure 3.** Illustration of the partitioning of microstates. Note, that in this step the transition probabilities on the microstate level are preserved.

The $\mathbf{P}_{ij}$ are block matrices and have the dimension $(m_i \times n_j)$, where $\sum_{i=1}^{\ell} m_i = \sum_{j=1}^{\ell} n_j = m$. Together with an approximation of the steady state eigenvector the block matrices are aggregated to form entries of a matrix $\mathbf{Q}$, which is used to solve the eigenvector problem on a smaller state space. The solution on the smaller state space is then disaggregated to improve the approximation of the steady state vector on the microstate space. The procedure is as follows:

(i) Choose the initial approximation of the steady state vector $\mathbf{x}^0 = (x_1^0, x_2^0, \ldots, x_m^0)$ and let $k = 1$.

(ii) Compute the vectors $\hat{\mathbf{x}}_i^{(k-1)}$ as

$$\hat{\mathbf{x}}_i^{(k-1)} = \frac{\mathbf{x}_i^{(k-1)}}{\left\|\mathbf{x}_i^{(k-1)}\right\|_1},$$

for $i = 1, \ldots, \ell$, $\left\|\mathbf{x}_i^{(k-1)}\right\|_1 = \mathbf{x}_i^{(k-1)}\mathbf{e}$ and $\mathbf{e}$ is a vector with all elements equal to one.

(iii) Obtain the aggregated matrix $\mathbf{Q}^{(k-1)}$ whose elements $q_{ij}^{(k-1)}$ are given by

$$\mathbf{q}_{ij}^{(k-1)} = \hat{\mathbf{x}}_i^{(k-1)}\mathbf{P}_{ij}\mathbf{e},$$

for $i, j = 1, \ldots, \ell$.[2]

(iv) Find the eigenvectors $\mathbf{w}^{(k-1)} = (w_1, w_2, \ldots, w_\ell)$ corresponding to the unit eigenvalue of matrix $\mathbf{Q}^{(k-1)}$, i.e. solve the problem

$$\mathbf{w}^{(k-1)}\mathbf{Q}^{(k-1)} = \mathbf{w}^{(k-1)}$$

and

$$\sum_{i=1}^{\ell} w_i^{(k-1)} = 1.$$

[2] Note, while the matrix $\mathbf{Q}$ represents the aggregated matrix, its entries are not to be confused with macrostate transition probabilities. They represent an aggregation of the iterative procedure $\mathbf{x}^{(k)} = \mathbf{x}^{(k-1)}\mathbf{P}$.

(v) Set up the diagonal matrix $\mathbf{Z}^{(k-1)}$ defined as

$$\mathbf{Z}^{(k-1)} = \mathrm{diag}\left\{ \frac{\mathbf{w}_1^{(k-1)}}{\|\mathbf{x}_1^{(k-1)}\|_1}\mathbf{I}_1, \frac{\mathbf{w}_2^{(k-1)}}{\|\mathbf{x}_2^{(k-1)}\|_1}\mathbf{I}_2, \ldots, \frac{\mathbf{w}_\ell^{(k-1)}}{\|\mathbf{x}_\ell^{(k-1)}\|_1}\mathbf{I}_\ell \right\},$$

where $\mathbf{I}_j$ are identity matrices of dimension $(n_j \times n_j)$.

(vi) Improve the approximation to the steady state vector via

$$\mathbf{x}^{(k)} = \mathbf{x}^{(k-1)}\mathbf{Z}^{(k-1)}\mathbf{K},$$

with

$$\sum_{i=1}^{m} x_i^{(k)} = 1 .$$

(vii) Check for convergence of the vector $\mathbf{x}^{(k)}$. This can be done by calculating $\eta = \|\mathbf{x}^{(k)} - \mathbf{x}^{(k-1)}\|$, where $\|\ldots\|$ is some norm. If $\eta < \epsilon$ the iteration procedure has reached its desired accuracy and can be exited, otherwise set $k = k + 1$ and continue at step (ii).

In step (vi) $\mathbf{K} = \mathbf{L}(\mathbf{D} - \mathbf{U})^{-1}$, where $\mathbf{L}$ and $\mathbf{U}$ are the lower triangular and the upper triangular matrix of the transition matrix $\mathbf{P}$, and $\mathbf{D} = \mathbf{I} - \mathrm{diag}\{\mathbf{P}\}$ is a diagonal matrix with the elements given by $d_{ii} = 1 - p_{ii}$ for $i = 1, \ldots, m$. $\mathbf{I}$ is the identity matrix and has dimension $(m \times m)$. $\mathbf{K}$ has to be calculated only once and does not need to be updated during the iteration procedure. Further, convergence of the method may be improved by implementing a pre- or post-smoothing procedure [6]. In the pre-smoothing procedure the matrix $\mathbf{P}$ is multiplied by itself, i.e. $\mathbf{P} \to \mathbf{P}^{s+1}$, where $s$ is the number of pre-smoothing steps and $s = 0$ corresponds to the method without pre-smoothing steps. The post-smoothing procedure is implemented by raising $\mathbf{K}$ to a power, i.e. $\mathbf{K} \to \mathbf{K}^{r+1}$, where $r$ denotes the number of post-smoothing steps and $r = 0$ corresponds to the method without post-smoothing steps.

## 3. Numerical Examples

In this section I will show two examples. In the first example the iterative aggregation and disaggregation method will be illustrated at the calculation of the steady state vector of a system for which the matrix $\mathbf{P}$ is known. This example is for didactical purpose, it will be used to illustrate steps (i) to (vii) of the procedure. In the second example the method will be illustrated at particles undergoing Brownian motion over a one-dimensional fluctuating potential. The fluctuations represent a dichotomous change of the potential.

### *3.1. Didactic example with known transition matrix*

Let the matrix $\mathbf{P}$ be a $(5 \times 5)$ matrix with the following entries

$$\mathbf{P} = \begin{pmatrix} 0.4 & 0.3 & 0 & 0.15 & 0.15 \\ 0.2 & 0.5 & 0.2 & 0.1 & 0 \\ 0.1 & 0.1 & 0.6 & 0.05 & 0.15 \\ 0 & 0.2 & 0.4 & 0.2 & 0.2 \\ 0.2 & 0 & 0 & 0.1 & 0.7 \end{pmatrix} .$$

Matrix $\mathbf{I} - \mathbf{P}$ decomposes into $\mathbf{L}$, $\mathbf{D}$ and $\mathbf{U}$ in the following way

$$\begin{pmatrix} 0 & 0 & 0 & 0 & 0 \\ 0.2 & 0 & 0 & 0 & 0 \\ 0.1 & 0.1 & 0 & 0 & 0 \\ 0 & 0.2 & 0.4 & 0 & 0 \\ 0.2 & 0 & 0 & 0.1 & 0 \end{pmatrix}, \begin{pmatrix} 0.6 & 0 & 0 & 0 & 0 \\ 0 & 0.5 & 0 & 0 & 0 \\ 0 & 0 & 0.4 & 0 & 0 \\ 0 & 0 & 0 & 0.8 & 0 \\ 0 & 0 & 0 & 0 & 0.3 \end{pmatrix} \text{ and } \begin{pmatrix} 0 & 0.3 & 0 & 0.15 & 0.15 \\ 0 & 0 & 0.2 & 0.1 & 0 \\ 0 & 0 & 0 & 0.05 & 0.15 \\ 0 & 0 & 0 & 0 & 0.2 \\ 0 & 0 & 0 & 0 & 0 \end{pmatrix} .$$

These are the lower triangular, the diagonal and the upper triangular matrix (respectively) of the decomposition $\mathbf{I}-\mathbf{P}=\mathbf{D}-\mathbf{L}-\mathbf{U}$, where $\mathbf{I}$ is the identity matrix. This gives for $\mathbf{K}=\mathbf{L}\left(\mathbf{D}-\mathbf{U}\right)^{-1}$

$$\begin{pmatrix} 0 & 0 & 0 & 0 & 0 \\ 0.3333 & 0.2 & 0.1 & 0.0938 & 0.2792 \\ 0.1667 & 0.3 & 0.15 & 0.0781 & 0.2104 \\ 0 & 0.4 & 1.2 & 0.125 & 0.6833 \\ 0.3333 & 0.2 & 0.1 & 0.2188 & 0.3625 \end{pmatrix} .$$

Do obtain the aggregated matrix $\mathbf{Q}$, the matrix $\mathbf{P}$ is partitioned into four submatrices $\mathbf{P}_{ij}$ (with $i,j=1,2$) in the following way

$$\begin{pmatrix} \begin{bmatrix} 0.4 & 0.3 & 0 \\ 0.2 & 0.5 & 0.2 \\ 0.1 & 0.1 & 0.6 \end{bmatrix} & \begin{bmatrix} 0.15 & 0.15 \\ 0.1 & 0 \\ 0.05 & 0.15 \end{bmatrix} \\ \begin{bmatrix} 0 & 0.2 & 0.4 \\ 0.2 & 0 & 0 \end{bmatrix} & \begin{bmatrix} 0.2 & 0.2 \\ 0.1 & 0.7 \end{bmatrix} \end{pmatrix} = \begin{pmatrix} \mathbf{P}_{11} & \mathbf{P}_{12} \\ \mathbf{P}_{21} & \mathbf{P}_{22} \end{pmatrix} .$$

The iterative procedure is then initiated according to step (i) by choosing a starting approximation $\mathbf{x}^0$ to the steady state vector. For this purpose, let $\mathbf{x}^0=(0.2\ \ 0.2\ \ 0.2\ \ 0.2\ \ 0.2)$, which is then partitioned into subvectors compatible with the partitioning of matrix $\mathbf{P}$ into the submatrices $\mathbf{P}_{ij}$, i.e.

$$\left(\begin{bmatrix}0.2 & 0.2 & 0.2\end{bmatrix} \ \begin{bmatrix}0.2 & 0.2\end{bmatrix}\right) = \left(\mathbf{x}_1^0 \ \ \mathbf{x}_2^0\right) .$$

Normalizing each of the subvectors gives for step (ii)

$$\left(\hat{\mathbf{x}}_1^0 \ \ \hat{\mathbf{x}}_2^0\right) = \left(\begin{bmatrix}\frac{1}{3} & \frac{1}{3} & \frac{1}{3}\end{bmatrix} \ \begin{bmatrix}\frac{1}{2} & \frac{1}{2}\end{bmatrix}\right) .$$

Note, the elements $\left(\hat{\mathbf{x}}_n^0\right)_{j_n}$ represent the approximation of the conditional probability of finding the system in one of the microstates $j_n$ given it is in the partition $n$ belonging to microstate $j_n$ (here $n=1,2$ denotes the partition and $j_1=1,\ldots,3$ and $j_2=4,5$ are the microstates). The aggregated matrix $\mathbf{Q}^0$ of step (iii) then becomes

$$\mathbf{Q}^0 = \begin{pmatrix} 0.8 & 0.2 \\ 0.4 & 0.6 \end{pmatrix} .$$

Solving the eigenvector problem on the aggregated state space, i.e. step (iv), gives the aggregated eigenvector

$$\mathbf{w}^0 = \left(\tfrac{2}{3} \ \ \tfrac{1}{3}\right) ,$$

and the diagonal matrix $\mathbf{Z}^0$ of step (v) becomes

$$\mathbf{Z}^0 = \begin{pmatrix} \frac{10}{9} & 0 & 0 & 0 & 0 \\ 0 & \frac{10}{9} & 0 & 0 & 0 \\ 0 & 0 & \frac{10}{9} & 0 & 0 \\ 0 & 0 & 0 & \frac{5}{6} & 0 \\ 0 & 0 & 0 & 0 & \frac{5}{6} \end{pmatrix} .$$

Note, here the entries in the diagonal matrix $\mathbf{Z}$ indicate that probability is transferred from the subvector $\hat{\mathbf{x}}_2^0$ to $\hat{\mathbf{x}}_1^0$. The disaggregation step (vi) gives for $\mathbf{x}^{(1)}$ after normalization

$$\mathbf{x}^{(1)} = \left(0.162 \ \ 0.2052 \ \ 0.2647 \ \ 0.0928 \ \ 0.2752\right) .$$

Implementing the Euclidian norm $\eta = \sqrt{\sum_i (x_i^{(k)} - x_i^{(k-1)})^2}$ for step (vii) gives

$$\eta = 0.15101 \, .$$

To reach a convergence better than $\epsilon = 10^{-5}$ the procedure has to be repeated eight more times, leading to

$$\mathbf{x}^{(9)} = \begin{pmatrix} 0.1959 & 0.2041 & 0.2123 & 0.1102 & 0.2775 \end{pmatrix} \, .$$

As comparison, implementing the standard iterative procedure $\mathbf{x}^{(k)} = \mathbf{x}^{(k-1)}\mathbf{P}$ requires eighteen iterations to reach the same convergence.

*3.2. Computational Example - Fluctuating Barrier*

To illustrate the full procedure, a one dimensional example is chosen where the dynamics represents an overdamped Brownian motion subject to a dichotomously changing potential [11], i.e.

$$\dot{x} = -\nabla \left[ u(x) + V_{DN}(\tau)x \right] + \Gamma(\tau) \, , \tag{5}$$

where $\eta$ is the friction constant, $\tau = \frac{k_B T}{\eta} t$, $u(x) = U(x)/k_B T$ and $\Gamma(\tau)$ is a white noise given by the following properties

$$\langle \Gamma(\tau) \rangle = 0$$
$$\langle \Gamma(\tau)\Gamma(\tau') \rangle = 2\delta(\tau - \tau') \, .$$

$V_{DN}$ represents a dichotomous noise alternating between the two fixed values

$$V^+ = \sqrt{\frac{D}{\tau_v}\left(\frac{1+\epsilon}{1-\epsilon}\right)} \qquad \text{and} \qquad V^- = -\sqrt{\frac{D}{\tau_v}\left(\frac{1-\epsilon}{1+\epsilon}\right)} \, ,$$

where $\tau_v$ is the correlation time of the dichotomous noise, $\epsilon$ a parameter introducing asymmetry and $\sqrt{D/\tau_v}$ the amplitude. The switching between the $V^+$ and $V^-$ state of the extended potential is correlated in time following

$$\langle V_{DN}(\tau) \rangle = 0$$

and

$$\langle V_{DN}(\tau)V_{DN}(0) \rangle = \frac{D}{\tau_v} \exp\left(-\frac{\tau}{\tau_v}\right) \, .$$

The potential $U(x)$ used in eq. (5) is Kramer‘s potential, i.e.

$$U(x) = -\frac{a}{2}x^2 + \frac{b}{2}x^4 \, .$$

The simulation is carried out using the predictor corrector algorithm for dynamics subject to dichotomous noise [12] with a time step $\mathrm{d}t = 3 \times 10^{-4}$. The parameters used are $\eta = 1$, $\epsilon = 0.8$, $a = 10$, $b = 1$, $k_B T = 0.15\Delta U$, where $\Delta U = a^2/4b$ is the height of the barrier between the two minima in Kramer‘s potential, and $D/\tau_v = 0.71$. A plot of the potential in its $V^+$ and $V^-$ state is shown in fig. (4).

The entries $p_{ij}$ in the matrix $\mathbf{P}$ are obtained using distributed computing. For this purpose the interval $[-6.5, 6.5]$ is divided into 30 intervals $\zeta_i$ of equal length [fig. (5) shows the first five intervals] and simulations are carried out individually for each of the inner states in the following way. For each independent run 500 trajectories are initiated in the interval $\zeta_i$ for $i = 2, \ldots, 29$ and allowed to evolve in the combined interval $\zeta_{i-1} \bigcup \zeta_i \bigcup \zeta_{i+1}$, which allows to

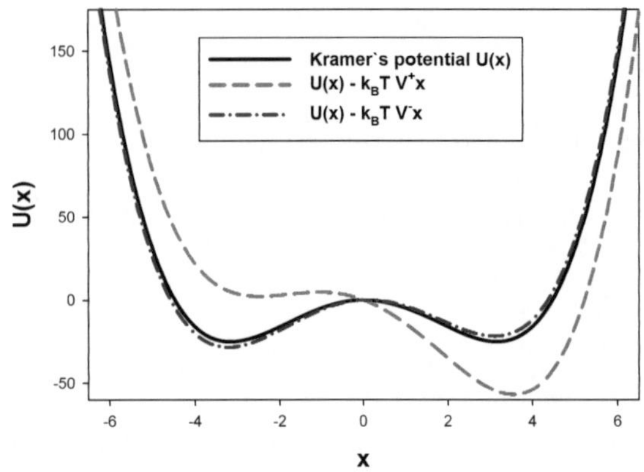


**Figure 4.** Plot of the effective potential in its $V^+$ (dashed line) and $V^-$ state (dash-dot line). The unaltered potential is drawn as solid line.

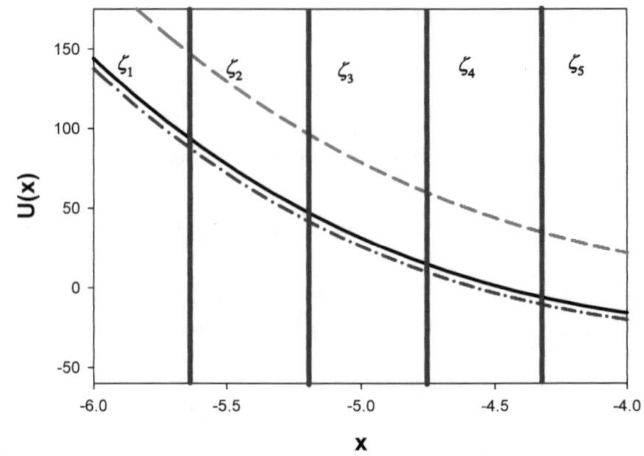


**Figure 5.** Illustration of the first five intervals $\zeta_1, \ldots, \zeta_5$ used to define the states. Note that these intervals were also used to define limited regions for distributed computing.

take into account local recrossing effects[3]. Trajectories which attempt to leave the combined interval via a transition from $\zeta_{i-1}$ to $\zeta_{i-2}$ (or from $\zeta_{i+1}$ to $\zeta_{i+2}$) are reflected at the boundaries of the simulation region back into $\zeta_{i-1}$ (or $\zeta_{i+1}$ respectively). After sufficient initial run, necessary to remove the effects of the starting location of the trajectories on the measurement of the transition probabilities, the single time step conditional probabilities $p(i-1|i)$ and $p(i+1|i)$ (for $i = 2, \ldots, 29$) of leaving state $\zeta_i$ to $\zeta_{i-1}$ and to $\zeta_{i+1}$ are measured using 10mio time steps.
The conditional probabilities are then used to set up the transition matrix $\mathbf{P}$ using the transition probabilities $p_{ij}$ according to eq. (3). Note, during the simulation it is observed that some of the states are not reached from other states. These states represent high energy outer states of the potential energy, and are cut out of the matrix $\mathbf{P}$. This leads to tri-diagonal Markov-state matrices of dimension $(25 \times 25)$ for the correlation times $\tau_v = 3 \times 10^{-3}, \ldots, 1 \times 10^{-2}$ and $\tau_v = 6 \times 10^{-1}, \ldots, 2.5 \times 10^{1}$ and to matrices of dimension $(26 \times 26)$ for $\tau_v = 2.5 \times 10^{-2}, \ldots, 2.5 \times 10^{-1}$ and $\tau_v = 6 \times 10^{1}, \ldots, 6 \times 10^{3}$.
For the case of 25 connected states, the matrix $\mathbf{P}$ is then partitioned into submatrices $\mathbf{P}_{ij}$ of dimension $(5 \times 5)$, leading to an aggregated matrix $\mathbf{Q}$ of dimension $(5 \times 5)$. For the case of 26 connected states, the elements $p_{ij}$ with $i, j = 1, \ldots, 15$ are partitioned into submatrices as before and the elements $p_{(26)j}$ with $j = 1, \ldots, 26$ and $p_{i(26)}$ with $i = 1, \ldots, 25$ are partitioned as $(1 \times 5)$- and $(5 \times 1)$-dimensional submatrices $\mathbf{P}_{6j}$ for $j = 1, \ldots 5$ and $\mathbf{P}_{i6}$ for $i = 1, \ldots 5$. The submatrix $\mathbf{P}_{66}$ is comprised of only one element, namely $p_{(26)(26)}$.
The iterative procedure is started [i.e. step (i)] with an equal distribution of the initial approximation $\mathbf{x}^0$, i.e. $x_i = 1/25$ ($\forall\ i = 1, \ldots, 25$) or $x_i = 1/26$ ($\forall\ i = 1, \ldots, 26$) respectively, and stopped once the norm $\eta = ||x^{(k)} - x^{(k-1)}||$ is smaller than $\epsilon = 1 \times 10^{-4}$. As norm the $L^2$ norm was chosen, i.e. the Euclidian norm where $\eta = \sqrt{\sum_i (x_i^{(k)} - x_i^{(k-1)})^2}$.

Figure (6) shows for different dichotomous noise correlation times $\tau_v$ a plot of the probability of steady state occupation of the left potential well (i.e. the region in space for which $x \leq 0$). The results for the distributed computing are compared to those from a full simulation using continuous trajectories. As can be seen from the figure, for fast switching times the results

[3] Local in the sense that it takes the configuration-space history of the trajectory into account only up to the neighboring states, but not farther.

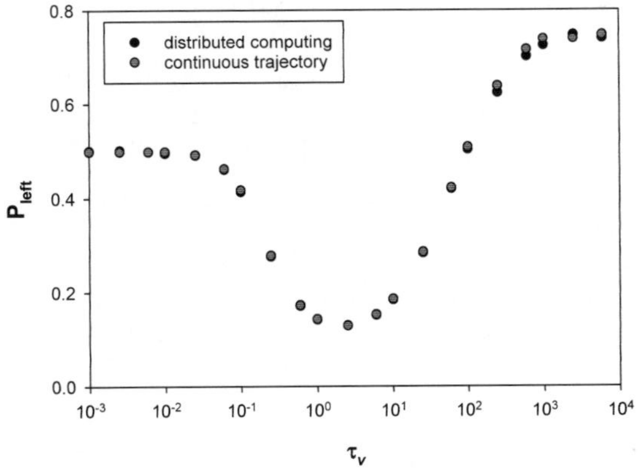


**Figure 6.** Steady state probability of occupation of the left potential well $\mathbf{P}_{\mathrm{left}}$ for different switching times $\tau_v$. The transition matrices were obtained using distributed computing (black dots) and continuous trajectories (red dots).

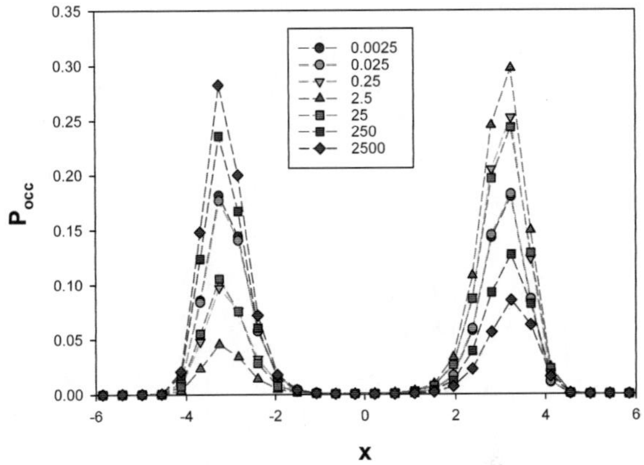


**Figure 7.** Steady state occupation of the states **x** for some of the switching times $\tau_v$.

are in agreement within the convergence bound of the iterative procedure. For slow switching times, i.e. $\tau_v > 10^2$, the difference between the values is bigger. This could be an indication of the effect the distributed computing procedure could have on the accuracy of the transition probabilities. Overall, the results agree with those reported by other authors [11]. A plot of the probabilities of occupation of the individual states is shown in fig. (7) for some of the correlation times $\tau_v$ (numbers shown in the legend). It can be seen from fig. (7) that at correlation times of $\tau_v = 2.5 \times 10^{-3}$ and $2.5 \times 10^{-2}$ the left and right potential well are approximately equally occupied, whereas for slow correlation times states corresponding to the left potential well have higher occupancy than states within the right potential well. The minimum of the occupancy of the left potential well is found around $\tau_v = 2.5$ [compare figure (6) and (7)], indicating that at this dichotomous noise correlation time nonequilibrium kinetic focusing occurs.

## 4. Summary

In this paper a method is introduced which combines the mapping from a continuous process on the phase- or configuration-space to a stochastic matrix representing the probabilities of transitions between an abstract set of states, with that of an iterative aggregation and disaggregation procedure in an effort to effectively obtain the steady state distribution of the process. The method is shown on two examples. The first example is of didactical nature, highlighting the different steps in the iterative procedure, whereas the second example shows a Brownian motion process subject to a dichotomously changing potential. Here the transition probabilities of the stochastic matrix were obtained using distributed computing and the resulting probabilities of occupation were compared to those from full trajectory simulations for different rates of fluctuation. For correlation times of the dichotomous noise $\tau_v \leq 10^2$ the results for the probability of occupation of the left potential well are in good agreement, within the convergence bounds, for both the procedures. Faster dichotomous noise correlation times show a larger difference between the steady state distributions, which possibly could indicate limitations of the distributed computing procedure and will be subject to further studies. Note, the use of transition probabilities rather than direct trajectory information makes the method suitable for the use for large scale bio-physical processes. In this paper the method was implemented using a simple distributed computing procedure, where the transition probabilities were obtained from independent runs of different domains, but could also be combined with other simulation procedures such as trajectory parallelization [13] or parallel replica dynamics [14].

## 5. References

[1] Bowman G, Pande V and Noé F 2014 *An Introduction to Markov State Models and Their Application to Long Timescale Molecular Simulation* (Dordrecht: Springer) and references therein
[2] Schütte C and Sarich M 2013 *Metastability and Markov State Models in Molecular Dynamics* (Providence: Courant Lecture Notes, AMS)
[3] Cao W L and Steward W 1985 *J. Assoc. Comput. Mach.* **32** 702
[4] Pultarová I and Marek I 2011 *Numer. Linear Algebra* **18** 1051
[5] Pultarová I and Marek I 2011 *J. Comput. Appl. Math.* **236** 354
[6] Biswas K 2015 *Comput. Phys. Commun.* **191** 25
[7] Ipsen I and Kirkland S 2006 *SIAM J. Matrix Anal. Appl.* **27** 952
[8] Sterck H, Manteuffel T, McCormic S, Nguyen Q and Ruge J 2008 *SIAM J. Sci. Comput.* **30** 2235
[9] Biswas K and Novotny M A 2011 *J. Phys. A: Math. Theor.* **44** 345004
[10] Biswas K 2013 *J. Phys. A: Math. Theor.* **46** 145001
[11] Ponzoni L, Celardo G, Borgonovi F, Kaplan L and Kargol A 2013 *Phys. Rev. E* **87** 052137
[12] Barik D, Ghosh P K and Ray D S 2006 *J. Stat. Mech.* **03** P03010
[13] Vanden-Eijnden E and Venturoli M 2009 *J. Chem. Phys.* **131** 044120
[14] Perez D, Uberuaga B P and Voter A F 2015 *Comput. Mater. Sci.* **100** 90

IOP Conf. Series: Journal of Physics: Conf. Series **921** (2017) 012012 doi:10.1088/1742-6596/921/1/012012

# High resolution Monte Carlo study of the Domb-Joyce model

**Nathan Clisby**

School of Mathematics and Statistics, The University of Melbourne, Victoria 3010, Australia

E-mail: nclisby@unimelb.edu.au

**Abstract.** We study the Domb-Joyce model of weakly self-avoiding walks on the simple cubic lattice via Monte Carlo simulations. We determine to excellent accuracy the value for the interaction parameter which results in an improved model for which the leading correction-to-scaling term has zero amplitude.

## 1. Introduction

The Domb-Joyce model of weakly self-avoiding walks has a long history [1]. It consists of the set of random walks on the simple cubic lattice $\mathbb{Z}^3$, with walks starting at the origin and taking steps to adjacent neighbors, where there is a contact interaction which introduces an energy penalty of $w$ for each pair of visits to the same lattice site. A walk of $N$ steps may be defined as a mapping $\omega$ from the integers $0, 1, \cdots, N$ to sites on $\mathbb{Z}^3$, with $|\omega(i+1)-\omega(i)| = 1 \; \forall i \in [0, N-1]$, and $\omega(i) \neq \omega(j) \; \forall i \neq j$, and the energy due to overlaps of a walk $\omega$ is then given by

$$E(\omega) = w \sum_{i<j} \delta_{\omega(i),\omega(j)}. \tag{1}$$

Note that if a walk visits a particular site exactly $k$ times, then the associated weight for that site will be equivalent to $\binom{k}{2}$ pairwise interactions. We then define the partition function as

$$Z_N = \sum_{|\omega|=N} e^{-E(\omega)}, \tag{2}$$

and the expectation of an observable $A$ is computed over the set of all simple random walks of length $N$ as follows:

$$\langle A \rangle_N = \frac{1}{Z_N} \sum_{|\omega|=N} A(\omega) e^{-E(\omega)}. \tag{3}$$

The Domb-Joyce model interpolates between simple random walks when $w = 0$, and self-avoiding walks when $w = +\infty$.

Here we perform Monte Carlo simulations of Domb-Joyce walks via the pivot algorithm [2,3] using a recent implementation [4,5] (which improved on earlier work by Kennedy [6]).

We calculated the expected value of the two most common measures of size, the squared end-to-end distance, $R_{\mathrm{E}}^2$, and the squared radius of gyration, $R_{\mathrm{G}}^2$, which are defined as:

$$R_{\mathrm{E}}^2 = |\omega(N) - \omega(0)|^2; \tag{4}$$

$$R_{\mathrm{G}}^2 = \frac{1}{2(N+1)^2} \sum_{i,j} |\omega(i) - \omega(j)|^2. \tag{5}$$

For any positive $w$ the Domb-Joyce model is in the same universality class as self-avoiding walks, and the asymptotic behavior of these observables for large $N$ is given by

$$\langle R_{\mathrm{E}}^2 \rangle = D_{\mathrm{E}}(w) N^{2\nu} \left( 1 + \frac{a_{\mathrm{E}}(w)}{N^{\Delta_1}} + \cdots \right), \tag{6}$$

$$\langle R_{\mathrm{G}}^2 \rangle = D_{\mathrm{G}}(w) N^{2\nu} \left( 1 + \frac{a_{\mathrm{G}}(w)}{N^{\Delta_1}} + \cdots \right). \tag{7}$$

The critical exponent $\nu$, known as the Flory exponent, is a universal quantity with value $\nu = 0.58759700(40)$ [7]. The leading correction-to-scaling exponent $\Delta_1$ is also universal, and the best estimate for it is $\Delta_1 = 0.528(8)$. The amplitudes $D_{\mathrm{E}}(w)$, $D_{\mathrm{G}}(w)$, $a_{\mathrm{E}}(w)$, and $a_{\mathrm{G}}(w)$ are non-universal quantities which depend on the details of the model, including the value of the weight $w$ and the lattice type. But, crucially for the present study, the amplitude ratios $D_{\mathrm{E}}(w)/D_{\mathrm{G}}(w) = 6.253531(10)$ [7] and $a_{\mathrm{E}}(w)/a_{\mathrm{E}}(w)$ are universal quantities which are independent of $w$.

A key reason for the continuing interest in the Domb-Joyce model is the observation that the addition of the interaction parameter $w$ allows for the possibility of tuning it to a value such that the leading correction-to-scaling terms have negligible amplitude. As $a_{\mathrm{E}}(w)/a_{\mathrm{E}}(w)$ is universal, if there is a value $w = w^*$ such that $a_{\mathrm{E}}(w^*) = 0$, then $a_{\mathrm{G}}(w^*) = 0$ also. Thus, if $w^*$ can be found to good accuracy it will ensure that the leading correction-to-scaling term for each observable will have small amplitude, thus enhancing convergence in the large-$N$ limit. The Domb-Joyce model with $w = w^*$ is called an "improved" model due to this feature.

Previously, Belohorec [8] estimated that $w^* = 0.506$ for the Domb-Joyce model, and Caracciolo et al. [9] found that $w^* = 0.48(2)$ when studying virial coefficients for the self-avoiding walk universality class. More recently, Adamo and Pelissetto [10] made a detailed study of the Domb-Joyce model in the presence of hard spheres, and obtained an estimate of $w^* = 0.486 \pm 0.003 \pm 0.005$, where the first error is statistical, and the second comes from varying parameters which were used to bias their fits.

Here we seek to build on earlier work and obtain an accurate estimate of $w^*$ to facilitate future Monte Carlo studies of properties of the self-avoiding walk universality class.

## 2. Monte Carlo simulation

We sampled self-avoiding walks on $\mathbb{Z}^3$ over a range of lengths from one thousand to ten million steps, for weights $e^{-w} = 0.50$, 0.54, 0.57, 0.59, 0.60, 0.61, 0.63, 0.66, and 0.70 ($w = 0.6931$, 0.6162, 0.5621, 0.5276, 0.5108, 0.4943, 0.4620, 0.4155, and 0.3567 respectively). We used a variant of the SAW-tree implementation of the pivot algorithm as described in [4] to collect data for the observables $\langle R_{\mathrm{E}}^2 \rangle$ and $\langle R_{\mathrm{G}}^2 \rangle$.

The SAW-tree implementation was adapted to count the number of intersections created when a pivot move is performed, whereas for self-avoiding walks one only needs to know if an intersection occurs. To initialize the system we used a variant of the pseudo_dimerize procedure, and to eliminate any initialization bias we then performed approximately $20N$ successful pivots before collecting any data.

IOP Conf. Series: Journal of Physics: Conf. Series **921** (2017) 012012 doi:10.1088/1742-6596/921/1/012012

We sampled pivot sites uniformly at random along the chain, and with the pivot symmetry operations sampled uniformly at random from amongst the 47 lattice symmetries of $\mathbb{Z}^3$ that do not correspond to the identity.

After initialization, we started collecting data for our observables for each time step, and aggregated the results in batches of $10^8$.

The computer experiment was run for 130 thousand CPU hours on Dell PowerEdge FC630 machines with Intel Xeon E5-2680 CPUs. (Run in hyperthreaded mode, so in fact 260 thousand thread hours were used.) In total there were $5.7 \times 10^5$ batches of $10^8$ attempted pivots, and thus there were a grand total of $5.7 \times 10^{13}$ attempted pivots across all walk sizes and values of the parameter $w$.

Our data for $\langle R_\mathrm{E}^2 \rangle$,$\langle R_\mathrm{G}^2 \rangle$, and $\langle R_\mathrm{E}^2 \rangle / \langle R_\mathrm{G}^2 \rangle$ for different values of $w$ are collected in Tables A1–A9 of [11].

## 3. Analysis

Finding a good method to analyze our data and obtain an accurate estimate for $w^*$ is a difficult problem. The heart of the difficulty is that we have conflicting goals that need to be accommodated, namely we wish to reduce the influence of unfitted corrections to scaling and thus go to the large-$N$ limit, and we wish to study the Domb-Joyce model in the vicinity of $w \approx w^*$, but at the same time we want a large signal for the leading-correction-to-scaling term, which necessitates collecting data for intermediate values of $N$, and for values of $w$ which are quite far from $w^*$.

The most robust and accurate method we could devise involves the use of information about the value of $D_\mathrm{E}/D_\mathrm{G} = 6.253531(10)$ from simulations of SAWs in the large-$N$ limit [7]. For fixed $N$, we found the value of $w(N)$ such that $\langle R_\mathrm{E}^2 \rangle_N / \langle R_\mathrm{G}^2 \rangle_N = D_\mathrm{E}/D_\mathrm{G}$ via a quartic fit. These fits are perfect, in the sense that the reduced $\chi^2$ value for the fits are approximately one indicating that the model accurately describes the data, and so we are confident that the resulting estimates for $w(N)$ are reliable. We found that the error bars for $w(N)$ became too large to be useful for $N \geq 10^6$. One can see the reason for this in Fig. 1: as $N$ increases, $N^{-\Delta_1}$ decreases, the amplitude of deviations from the limiting value of $D_\mathrm{E}/D_\mathrm{G}$ become smaller, and so errors on the estimates for $w(N)$ increase.

To gain an understanding of the convergence behavior of our estimates $w(N)$ we examine the equation we are solving when the next correction-to-scaling term with exponent $-1$ is included. The expected asymptotic form for the ratio $\langle R_\mathrm{E}^2 \rangle_N / \langle R_\mathrm{G}^2 \rangle_N$ is

$$\frac{\langle R_\mathrm{E}^2 \rangle_N}{\langle R_\mathrm{G}^2 \rangle_N} = \frac{D_\mathrm{E}}{D_\mathrm{G}} \left( 1 + \frac{f(w)}{N^{\Delta_1}} + \frac{g(w)}{N} + \cdots \right). \tag{8}$$

By solving $\langle R_\mathrm{E}^2 \rangle_N / \langle R_\mathrm{G}^2 \rangle_N = D_\mathrm{E}/D_\mathrm{G}$ at finite $N$, we find the value of $w$ which forces the correction-to-scaling terms to be zero in aggregate, i.e. we are solving

$$\frac{f(w)}{N^{\Delta_1}} + \frac{g(w)}{N} = 0, \tag{9}$$

neglecting higher order corrections. Now, $N$ is large, and we expect the solution for $w$ to be close to $w^*$, and so we take $w = w^* + \Delta w$, and drop subleading terms:

$$f(w^* + \Delta w) = -g(w^* + \Delta w) N^{\Delta_1 - 1}, \tag{10}$$

$$f(w^*) + \Delta w f'(w^*) = -g(w^*) N^{\Delta_1 - 1}, \tag{11}$$

$$\Delta w = -\frac{g(w^*)}{f'(w^*)} N^{\Delta_1 - 1}. \tag{12}$$

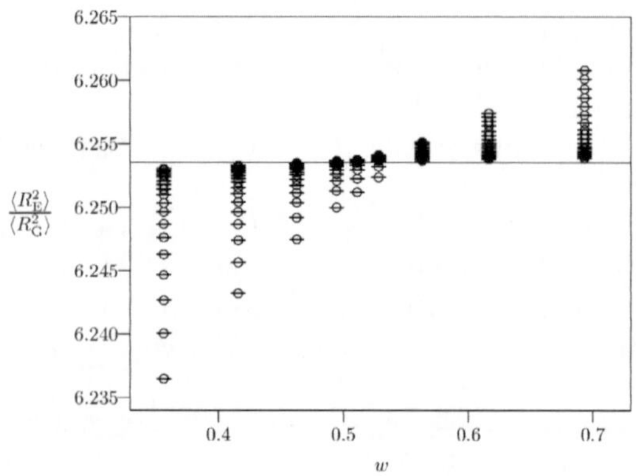


**Figure 1.** $\langle R_{\rm E}^2\rangle / \langle R_{\rm G}^2\rangle$ as a function of $w$ for $1000 \leq N \leq 10^7$. The smaller values of $N$ have larger corrections to scaling, and so deviate to a greater extent from the limiting value. We show a horizontal line for the best estimate of $D_{\rm E}/D_{\rm G} = 6.253531$.

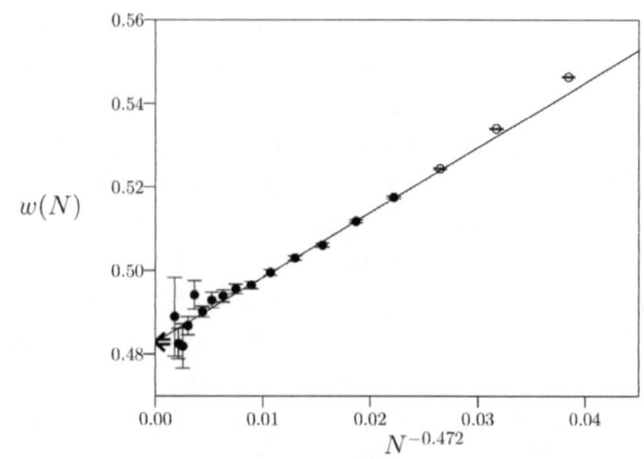


**Figure 2.** Estimates for $w(N)$ for $1000 \leq N \leq 680000$ plotted against $N^{\Delta_1 - 1}$ with $\Delta_1 = 0.528$. The estimate for $w^* = 0.48284(58)$, which is obtained by performing a linear fit of the solid data points, is shown on the vertical axis.

Thus we expect the deviation of our estimates $w(N)$ from the limiting value $w^*$ to be $\Delta w = O(N^{\Delta_1 - 1})$. One quite significant issue with this assumption is that there are competing next-to-leading correction terms, namely the analytic $O(N^{-1})$ correction term, the $O(N^{-2\Delta_1})$ correction term, and a further non-analytic correction term $O(N^{-\Delta_2})$ for which it is believed that $\Delta_2 \approx 1$. We will assume for now that the deviation is well described by $\Delta w = O(N^{\Delta_1 - 1})$, but will touch on this point again later in Sec. 4.

Now that we have $w(N)$, we plot the estimates against $N^{\Delta_1 - 1}$ with $\Delta_1 = 0.528$ in Fig. 2. We performed linear fits of these data, finding that the fit is excellent for $N$ in the range $3200 \leq N \leq 680000$, with reduced $\chi^2$ approximately one. The resulting estimate of $w^* = 0.48284(58)$ is shown in the plot on the vertical axis. Note that the estimates here for $w(N)$ are independent of each other, and hence it makes sense to perform fits and then derive a statistical error estimate for $w^*$, in contrast to, say, estimates of $\Delta_1$ in Fig. 10 of [7] which form a correlated sequence for which a linear fit no longer gives a meaningful statistical error.

Our estimate for $w^*$ is biased via the choice of values for $D_{\rm E}/D_{\rm G}$ and $\Delta_1$. Although we have only shown estimates in Fig. 2 which are biased with the central values of estimates for $D_{\rm E}/D_{\rm G}$ and $\Delta_1$, we have varied these values within their confidence intervals. We find that varying $D_{\rm E}/D_{\rm G} = 6.253531(10)$ within the confidence interval $\pm 0.000010$ causes a variation in the estimate of $w^*$ of $\pm 0.0011$. Varying $\Delta_1 = 0.528(8)$ within the confidence interval $\pm 0.008$ causes a variation in the estimate of $w^*$ of $\pm 0.00034$. Thus the largest source of error comes from our earlier estimate of $D_{\rm E}/D_{\rm G}$.

We combine the statistical error of 0.00058 with the errors due to biasing as if they were statistically independent sources of error, giving $\sigma(w^*) = \sqrt{0.00058^2 + 0.0011^2 + 0.00034^2} = 0.0013$. Thus our final estimate, incorporating all known sources of error, is $w^* = 0.4828(13)$.

Finally, we note that for the analysis we essentially have to take 2 limits, $N \to \infty$, and $w \to w^*$. The previously described method involves first solving for $w(N)$ (taking the limit $w \to w^*$ at finite $N$), and then taking the limit $w^* = \lim_{N\to\infty} w(N)$. We also tried taking the limit in the reverse order as follows. We fitted data for $\langle R_{\rm E}^2\rangle$ and $\langle R_{\rm G}^2\rangle$ with biased values for $\nu$ and $\Delta_1$, obtaining sequences of estimates for $a_{\rm E}(w)$ and $a_{\rm G}(w)$ which were their limiting (large $N$) values. We then solved the equations $a_{\rm E}(w^*) = 0$ and $a_{\rm G}(w^*)$ for $w^*$. Unfortunately, we found that this process gave significantly larger error bars, and so we will not report the details.

## 4. Discussion and conclusion

We hope that an ingenious method of analysis may yet be found which can extract more information from the data given in Tables A1–A9 of [11]. We note that a more accurate value for $D_E/D_G$ would result in a correspondingly more accurate estimate for $w^*$; it may be the case that such an improved estimate comes from a simulation of the Domb-Joyce model with the value for $w^*$ reported in this paper!

As mentioned in the previous section, the error bar for our estimate of $w^*$ relies on the fact that the next-to-leading correction term is well approximated by a single $O(N^{-1})$ term, despite the fact that there may be competing terms. The observation that the data in Fig. 2 is well described by a straight line gives us some confirmation that this is true, but nonetheless we are mindful that the competing terms may result in a subtle error that could be larger than our reported confidence interval for $w^*$. The only way we can think of to mitigate this possible source of error is to obtain better data for larger $N$, which would allow for easier extrapolation. But, this is a hard problem: statistical errors are relatively larger for large $N$, so it is unclear how much the situation would be improved by doing this.

The use of the improved version of the Domb-Joyce model is undoubtedly an attractive choice for many problems in polymer physics as it allows for faster convergence. We note that there is another alternative to improve convergence for particular observables by reducing the amplitude of the leading correction-to-scaling term. This has been done for the Ising model [12, 13], and more recently for self-avoiding walks [7, 14]. The manner in which the amplitude of the leading correction-to-scaling term is reduced is independent between the two methods, and so in fact it is possible to use both methods simultaneously and compound the effect. Thus, a study of the Domb-Joyce model at $w = w^*$ for the improved observable $R^2_{\text{imp}} = R^2_{\text{E}} - 4.478R^2_{\text{G}}$ [7] would be expected to have a completely negligible leading correction-to-scaling term.

Finally, we give our best estimate for the value of $w$ which gives an improved model as $w^* = 0.4828(13)$; we hope that it proves useful in future studies of the venerable Domb-Joyce model.

## Acknowledgements

Support from the Australian Research Council under the Future Fellowship scheme (project number FT130100972) and Discovery scheme (project number DP140101110) is gratefully acknowledged.

## References

[1] Domb C, Barrett A J and Lax M 1973 *J. Phys. A: Math., Nucl. Gen.* **6** L82
[2] Lal M 1969 *Mol. Phys.* **17** 57–64
[3] Madras N and Sokal A D 1988 *J. Stat. Phys.* **50** 109–186
[4] Clisby N 2010 *J. Stat. Phys.* **140** 349–392
[5] Clisby N 2010 *Phys. Rev. Lett.* **104** 055702
[6] Kennedy T 2002 *J. Stat. Phys.* **106** 407–429
[7] Clisby N and Dünweg B 2016 *Phys. Rev. E* **94** 052102
[8] Belohorec P 1997 *Renormalization group calculation of the universal critical exponents of a polymer molecule* Ph.D. thesis University of Guelph
[9] Caracciolo S, Mognetti B M and Pelissetto A 2006 *J. Chem. Phys.* **125** 094903
[10] Adamo G D and Pelissetto A 2016 *J. Phys. A.: Math. Theor.* **49** 504006
[11] Clisby N 2017 High resolution Monte Carlo study of the Domb-Joyce model (*Preprint* 1705.01249) URL http://arxiv.org/abs/1705.01249
[12] Hasenbusch M, Toldin F P, Pelissetto A and Vicari E 2007 *J. Stat. Mech.* **2007** P02016
[13] Hasenbusch M 2010 *Phys. Rev. B* **82** 174433
[14] Clisby N 2017 Scale-free Monte Carlo method for calculating the critical exponent $\gamma$ of self-avoiding walks (*Preprint* 1701.08415) URL http://arxiv.org/abs/1701.08415

# Application of artificial neural networks to identify equilibration in computer simulations

**Mitchell H Leibowitz, Evan D Miller, Michael M Henry, Eric Jankowski**

Micron School of Materials Science and Engineering, Boise State University, 1910 University Dr., Boise, Idaho 83725, United States

E-mail: `ericjankowski@boisestate.edu`

**Abstract.** Determining which microstates generated by a thermodynamic simulation are representative of the ensemble for which sampling is desired is a ubiquitous, underspecified problem. Artificial neural networks are one type of machine learning algorithm that can provide a reproducible way to apply pattern recognition heuristics to underspecified problems. Here we use the open-source TensorFlow machine learning library and apply it to the problem of identifying which hypothetical observation sequences from a computer simulation are "equilibrated" and which are not. We generate training populations and test populations of observation sequences with embedded linear and exponential correlations. We train a two-neuron artificial network to distinguish the correlated and uncorrelated sequences. We find that this simple network is good enough for $> 98\%$ accuracy in identifying exponentially-decaying energy trajectories from molecular simulations.

## 1. Introduction

Computer simulations that model physical systems in equilibrium are tools that have been essential to understanding an enormous variety of phenomena ranging from liquid argon[1], the self-assembly of colloids[2], criticality in spin glasses[3, 4], lipid self-assembly[5], DNA-origami[6], protein folding[7, 8], and the segregation of neighborhoods[9]. For the modeling of material systems, Metropolis Monte Carlo (MC)[10] and molecular dynamics (MD)[11] are the two most popular tools for sampling equilibrium ensembles of microstates. Both techniques generate sequences of microstates and the main cost of performing an MD or MC simulation is waiting for this sequence to converge to one that is representative of the thermodynamic ensemble at equilibrium, and then sampling enough microstates to perform ensemble averages with high precision. Sampling the equilibrium distribution of microstates at a particular thermodynamic state point is challenging because both MD and MC techniques are inefficient at overcoming free energy barriers that separate local minima in the multidimensional configurational landscape[12]. Consequently, the observables for a simulation (e.g. potential energy, order parameters) may appear to converge about a stable average for "long" times before relaxing to another stable average associated with lower free energy (Fig. 1). Every researcher performing MD or MC simulations has encountered this equilibration problem when asking themselves "Have I run my simulation for long enough?"

There are two correct answers to this question that are both unsatisfying. The first answer is "No." It is unsatisfying to know that the only way to be sure that the sampled microstates are

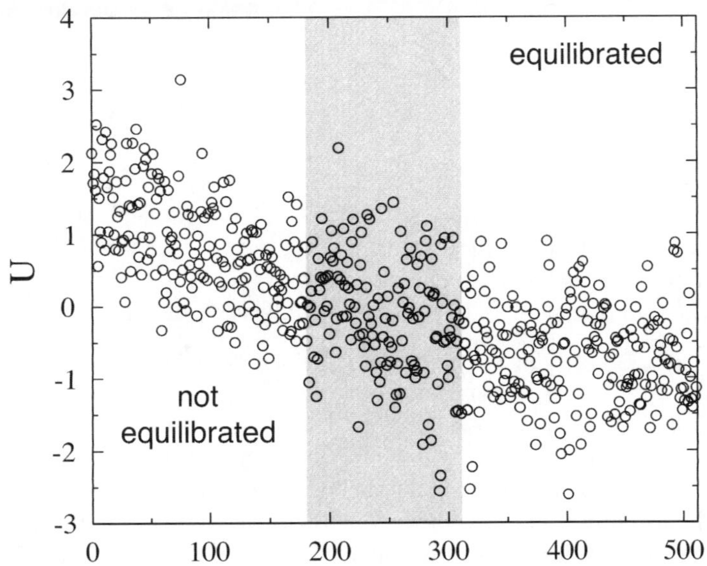


**Figure 1.** An exponential decay to a stable average is a frequently observed trend in thermodynamic simulations. It is difficult to decide where the exponential decay ends, and where the "equilibrated" region begins. Different heuristics could justify equilibrium sampling beginning anywhere in the shaded grey region.

representative of the free energy minimum is to perform a longer simulation that does not relax to a more stable distribution. That is, only in the limit of an infinitely long simulation can one really be sure that is has been run for long enough. The second answer is "Yes, when there are enough measurements to give a sufficiently precise answer." This answer is unsatisfying because it raises a new question "How shall I quantify when there are enough measurements?" and because it ignores the problem of whether the global free energy minimum has been sampled.

Recent advances in sampling schemes and computational hardware help to ameliorate, but not solve, the problem of generating infinitely many microstates. For MD simulations, the application of transition state theory[13] and histogram reweighting[14] has informed acceleration schemes that better sample rough energy landscapes. In MC simulations, enhanced sampling with Wang-Landau algorithms[3], cluster moves[15], and population annealing[16] are among the techniques used to extract more information per microstate. For both MD and MC simulations, accelerated sampling has been enabled with graphics processing units (GPUs) that facilitate single instruction-multiple data (SIMD) parallelization[17, 18].

Here we work towards addressing the second question "How shall I quantify when there are enough measurements." In principle, answering this question can be addressed by performing autocorrelation analysis on the system observables. In practice, measuring the autocorrelation time of system observables relies upon heuristics for what data to ignore or include. Even when the global free energy minimum is achieved after beginning with an initial random guess, which is a best-case scenario, the observed system energy decays exponentially from the initial guess to the correct average. In this case the problem becomes identifying the first microstate $u^*$ at time step $t^*$ after which the sequence of microstates $u_{t>t^*}$ represents the equilibrium distribution, using the distribution of observables such as system energy $U_t$ as a proxy for the distribution of microstates. In order to be reproducible, the process of identifying $u^*$ should be well-defined so that the same identification algorithm can be applied to all of the simulation trajectories generated for a scientific study, and to compare trajectories between studies. The strategy we use here to construct a well-defined heuristic for identifying $u^*$ is to train an artificial neural network for automating the decision of whether a sequence of observables is representative of an equilibrium distribution or not.

## 2. Methods

### *2.1. Measuring equilibration*

The $N$ microstates sampled during MC and MD simulations represent observations of the underlying, unknown equilibrium distribution. Averages computed from these independent observations should converge to a normal distribution, according to the central limit theorem. Therefore the three key metrics used to determine equilibration are the average $\langle U \rangle = \frac{1}{N}\sum_{i=1}^{N} U_i$, standard deviation $\sigma_U = \sqrt{\frac{1}{N}\sum_{i=1}^{N}(U_i - \langle U \rangle)^2}$, and decorrelation time $\delta t_0$ of a sequence of measurements. The autocorrelation function $a_U(\delta t)$ describes how correlated a sequence of measurements is with itself over a measurement lag $\delta t$ (steps in the case of MC, time in the case of MD), and the decorrelation time $\delta t_0$ where $a_U(\delta t_0) = 0$ describes how long one must wait, on average, for two measurements from the sequence $U_t$ to be decorrelated (independent) of each other. The ubiquity of fast numerical libraries for most programming languages makes computational implementations of the autocorrelation function relatively straightforward through the use of fast Fourier transforms:

$$a_U(\delta t) = \frac{\mathrm{IFT}[\mathrm{FT}[U_t - \langle U \rangle]\mathrm{FT}[U_t - \langle U \rangle]^*]}{N\sigma_U^2} \tag{1}$$

where FT$[A]$ is the discrete Fourier transform of arbitrary sequence $A$, IFT$[A]$ is the corresponding inverse Fourier transform, and FT$[A]^*$ is the complex conjugate of the Fourier transform of $A$. In python, this can be accomplished in a few lines of code:

```
import numpy
FT = numpy.fft.rfft #discrete Fourier transform on real input
IFT = numpy.fft.irfft #inverse discrete Fourier transform on real input
def autocorr1D(array):
    ft = FT(array-numpy.average(array))
    acorr = IFT(ft*numpy.conjugate(ft))/(len(array)*numpy.var(array))
    return acorr[0:len(acorr)//2]
```

where only the first half of the autocorrelation array is returned because of its symmetry with the second half. The above normalization by the variance $\sigma_U^2$ and sample size $N$ sets $a_U(0) = 1$, which indicates the fact that every measurement is perfectly correlated with itself (lag $\delta t = 0$). Common heuristics for determining $\delta t_0$ from $a_U(\delta t)$ include finding the smallest $\delta t$ where $a_U(\delta t) < 0$ (this avoids having to interpolate between values of $a_U(\delta t)$ to find the precise zero), or finding the fist $\delta t$ within some tolerance of 0.

Using $\langle U \rangle$, $\sigma_U^2$, and $\delta t_0$, the process of measuring equilibration can be thought of as finding the largest subset of the sequence $U_t$ that will give "enough" independent samples $N/\delta t_0$, specified by the researcher. Herein lies the challenge of specifying a well-defined heuristic, because it can depend on what one considers to be "enough", the frequency with which $U_t$ is written out by a simulation engine, and the relaxation kinetics for the system being studied, and the details of the scheme used to advance from $U_i$ to $U_{i+1}$. An example of one heuristic is to split $U_t$ into subsets of size $n$, measuring $\langle U \rangle$, $\sigma_U^2$, and $\delta t_0$ for each subset to determine if they are significantly different (e.g. with the student's t-test). The final equilibrated ensemble in this case is determined by concatenating the subsets moving backwards in time until one is found to be significantly different.

### *2.2. Machine Learning*

Another way to think about the problem of measuring equilibrium is to frame it as a pattern recognition problem, for which machine learning algorithms are well-suited[19]. A recent application of deep neural networks to pattern recognition in the board game *go* represents a breakthrough in computational *go* strategy[20]. Neural networks trained to recognize "good" moves were used to simplify the search space of Monte Carlo tree estimates of move quality[21] to beat over fifty professional go players in 2016 and early 2017. In the field of molecular simulation, machine learning techniques have been applied to identify assembly pathways in colloidal systems[2, 22], optimizing coarse-graining strategies[23], structural analysis[24], and design rules for organic photovoltaic ingredients[25]. Here we investigate the automated evaluation of equilibrated simulation trajectories using a single artificial neural network.

An artificial neural network is described by a set of "neurons" $n_i$ and "connections" $c_{i,j}$ between them. Each neuron "fires", generating an output signal that depends on the sum of its input connections, and the output signals can be used as input to more neurons or interpreted as decisions for pattern recognition problems. Artificial neural networks can be "trained" by determining the connection weights that minimize the error for a pattern recognition problem.

Here we specify a single "layer" of a neural network with two neurons. The first neuron is the "equilibrated" neuron and should fire when its input connections correspond to signal from an equilibrated sequence of observations. The second neuron is the "not equilibrated" neuron and should fire when its input connections correspond to signal from a non-equilibrated sequence of observations. Each neuron has one input connection for each observation and only one output connection. We can therefore specify this neural network as a matrix multiplication problem:

$$\begin{bmatrix} w_{1,1} & w_{1,2} & \cdots & w_{1,N} \\ w_{2,1} & w_{2,2} & \cdots & w_{2,N} \end{bmatrix} \cdot \begin{bmatrix} U_1 \\ U_2 \\ \vdots \\ U_N \end{bmatrix} + \begin{bmatrix} b_1 \\ b_2 \end{bmatrix} = \begin{bmatrix} n_1 \\ n_2 \end{bmatrix} \tag{2}$$

where the weights $w_{i,j}$ and biases $b_{i,j}$ are scalars that will be determined through a training process. The weights $w_{1,1\cdots N}$ correspond to the connection weights into the first neuron and $w_{2,1\cdots N}$ for the second neuron. After being trained, the artificial neural net defined by $w_{i,j}$ and $b_i$ can be used to determine if a sequence of $N$ observations is equilibrated by using $U_{1\cdots N}$ as input to Eqn. 2, resulting elements of a decision vector $n_1$ and $n_2$. During training we impose constraints to ensure $n_1 + n_2 = 1$, so the values returned by the artificial neural net can be interpreted as the certainty with which it thinks a data set is equilibrated or not.

### *2.3. Training the Neural Net*

To determine the $w_{i,j}$ and $b_i$ that most accurately distinguish equilibrated from non-equilibrated observation sequences we generate a set of $N_t$ training sequences with equal numbers of equilibrated and non-equilibrated samples. The equilibrated samples are generated using the Mersenne Twister to produce pseudorandom numbers normally distributed about 0. Two types of non-equilibrated samples are generated; linear and exponentials. The linear non-equilibrated samples are generated with slopes varying from -0.1 to -1.1 and with the same type of normally distributed noise about the net slope. The exponential non-equilibrated samples add $\exp(-\tau x)$ to the equilibrated case, for $1.5 \leq \tau \leq 3$. Sample python code for generating these populations in included below:

```
import numpy as np
def eq(n=512):
    return np.random.randn(n)
```

```
def neq_linear(n=512,min_slope=0.1):
    slope = -(np.random.rand()+min_slope)
    return np.arange(n)*slope/n+np.random.randn(n)

def neq_exp(n=512):
    e = np.exp(-3*(np.random.rand()+1.)*np.arange(n)/2./n)
    return e+np.random.randn(n)

%Generate training data with both types of non-equilibrium functions
def training_data(n_samples=2000,size=512):
    plots = eq(size)
    labels = np.array([1.0,0.0])
    for i in range(n_samples//2-1):
        plots = np.vstack((plots,eq(size)))
        labels = np.vstack((labels,np.array([1.0,0.0])))
    for i in range(n_samples//2):
        func = neq_linear
        if i%2==0:
            func = neq_exp
        plots = np.vstack((plots,func(size)))
        labels = np.vstack((labels,np.array([0.0,1.0])))
    return plots, labels
```

Before training the neural net on these sample observable sequences, we normalize them by subtracting out the sequence's average and dividing by its standard deviation. This normalization helps to ensure the neural net is trained to recognize relative changes in sequence for sequences of arbitrary magnitude. Each data set is produced with 64, 128, 256, or 512 x-values on the interval $[0, 1)$. The exponential decay constants and linear slopes used here are justified by being representative of simulation data and impossible for the authors to distinguish by eye (Fig. 2). Autocorrelation analysis of the equilibrated data sets gives $\delta t_0 = 1$ (neighboring samples are perfectly decorrelated, as expected), while the linear (slope $m = -0.1$) and exponential (decay time $\tau = 3$) nonequilibrated samples have $\delta t_0 = 3 \pm 1$. Here we train the neural net and charachterize the degree to which an it can distinguish the differences between these representative sequences.

We perform experiments with training populations of $N_t \in \{1600, 3200, 8000\}$, and generate independent test populations that are $\frac{1}{4}$ the size of the training population. We employ the TensorFlow python package for machine learning to facilitate training and testing[26]. TensorFlow can also take advantage of installed GPUs to accelerate its internal routines. Explaining the framework and syntactical details of TensorFlow is beyond the scope of this work, and can be found in online tutorials, including `www.tensorflow.org`. The following 18 lines of python use TensorFlow to define the network inputs and outputs as in Eqn. 2, the objective function that is minimized during training, perform the training, perform testing, and then returns the degree of accuracy ($[0, 1]$) with which the trained network correctly predicted the test data:

```
import tensorflow as tf
def single(train_plots, train_labels,test_plots,test_labelsn=512,rounds=2000):
    x = tf.placeholder(tf.float32,[None,n])
    W = tf.Variable(tf.zeros([n,2]))
    b = tf.Variable(tf.zeros([2]))
    y = tf.matmul(x,W)+b #Eqn. 2
    y_ = tf.placeholder(tf.float32,[None,2])
```

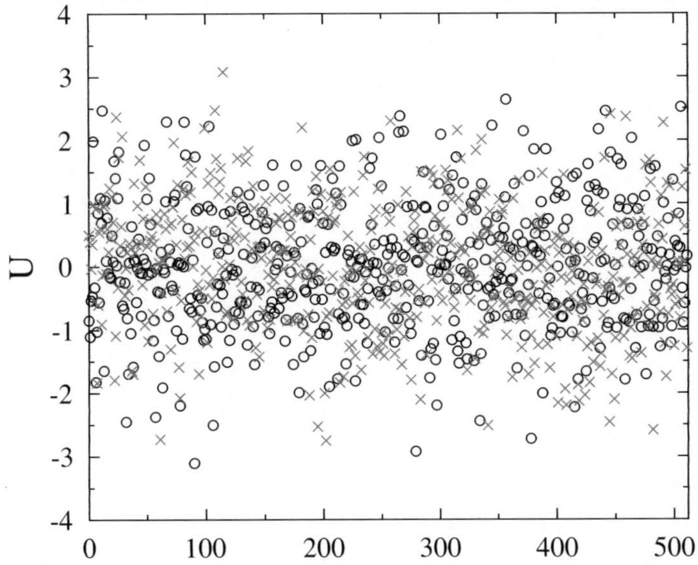


**Figure 2.** Normally-distributed observations (black) and normally-distributed observations with $\exp(-3x/512)$ added (red). The artificial neural network tested here can distinguish these data sets over 96% of the time.

```
obj = tf.nn.softmax_cross_entropy_with_logits(labels=y_, logits=y)
cross_entropy = tf.reduce_mean(err)
train_step = tf.train.GradientDescentOptimizer(0.5).minimize(cross_entropy)
sess = tf.Session()
init = tf.global_variables_initializer()
sess.run(init)
for _ in range(rounds): #do the training
    sess.run(train_step, feed_dict={x:train_plots,y_:train_labels})
correct = tf.equal(tf.argmax(y,1),tf.argmax(y_,1))
accuracy = tf.reduce_mean(tf.cast(correct,tf.float32))
return sess.run(accuracy,feed_dict={x:test_plots,y_:test_labels})
```

Here we investigate how information quality in the form of sample size $N$, number of training samples $N_t$, number of training rounds $r$ for a single layer artificial neural net influences is predictive accuracy $A$ for sample equilibrated and non-equilibrated observation sequences.

## 3. Results

Table 1 summarizes the predictive accuracy of the neural network as a function of information quality, for the training case where linear and exponential non-equilibrated data sets are included in equal parts, and where the non-equilibrated data has randomly generated slopes and decay rates, respectively. As is expected, predictive accuracy increases with the size of the input sequence $N$ and the number of training samples $N_t$. The number of training rounds (1000,2000,4000) is found to weakly influence accuracy. To better understand which observation sequences this neural net is best suited to distinguish, we restrict the types of non-equilibrium sequences are included in the training and testing sets and perform additional experiments. In *small* experiments we use $N = 256$, $N_t = 1600$, $r = 1000$, and for *large* experiments we use $N = 512$, $N_t = 8000$, $r = 2000$. The small experiments take less than thirty seconds on an

**Table 1.** Prediction accuracy when randomly sloped linear and randomly decaying exponential plots are included in the training and test population samples.

| $N$ | $N_t$ | $r$ | $A$ |
|---|---|---|---|
| 64 | 1600 | 2000 | .79 |
| 128 | 1600 | 2000 | .82 |
| 256 | 1600 | 2000 | .84 |
| 512 | 3200 | 4000 | .86 |
| 512 | 3200 | 1000 | .88 |
| 512 | 1600 | 2000 | .89 |
| 512 | 3200 | 2000 | .90 |
| 512 | 8000 | 1000 | .91 |

NVIDIA GeForce GT 750M GPU, while the large experiments take up to three minutes. We distinguish the accuracies that correspond to these tests with $A_s$ and $A_l$ respectively.

In the case of only linear sequences with constant slope $m = -0.1$, we find the network is unable to reliably distinguish these from equilibrated data ($A_s = .566$ and $A_l = .599$). If the slope of the linear sequences is increased to $m = -.2$ we find the network does relatively better ($A_s = .664$ and $A_l = .716$), suggesting that the network can detect steeper slopes more easily. When the slope of the linear sequences is uniformly chosen on $[0.2, 2.2)$ we find the network does even better ($A_s = .862$ and $A_l = .912$).

We find the network is more effective at distinguishing randomized exponentially decaying observations than equilibrium observations, with $A_s = .968$ and $A_l = .996$ for $exp(-\tau x)$ and $\tau = -3$. When $\tau$ is uniformly chosen on $[1.5, 3)$, we find the network is approximately as predictive as in the constant $\tau$ case, with $A_s = .968$ and $A_l = .990$.

## 4. Conclusions

Artificial neural networks are a promising advance for automating the identification of "equilibrated" measurements from thermodynamic simulations. A few minutes of training time on a laptop GPU is sufficient to train a two-neuron artificial network for identifying an exponential decay hidden in normally-distributed data more than 96% of the time. As this is the simplest possible artificial neural net that can be applied to the equilibrium-or-not decision problem, this suggest that neural networks with multiple layers of neurons can be applied to this problem with greater efficacy. The high accuracy of identifying random sequences with exponential decay is encouraging, because these sequences are characteristic of relaxations to equilibrium in thermodynamic simulations. However, our artificial neural network was not able to reliably distinguish random sequences with a small, constant slope embedded within them. Improvements to the two-neuron network studied here are needed for more robust pattern recognition. Future work applying multilayered artificial neural networks to automatically identify the subset of observations from a sequence that best represent equilibration are warranted, and seem accessible given the successes of deep (thousands of layers) networks in more computationally demanding pattern recognition problems. Open source machine learning libraries including TensorFlow provide a way to apply pattern recognition to problems in computer simulation, making these future studies more accessible, or even routine.

## References

[1] Rahman A 1964 *Phys. Rev. A* **136** A405–A411

IOP Conf. Series: Journal of Physics: Conf. Series **921** (2017) 012013 doi:10.1088/1742-6596/921/1/012013

[2] Long A W and Ferguson A L 2014 *J. Phys. Chem. B* **118** 4228–4244
[3] Wang F and Landau D P 2001 *Phys. Rev. Lett.* **86** 2050–2053
[4] Perera D, Vogel T and Landau D P 2016 *Phys. Rev. E* **94** 043308 2470-0045
[5] Gai L, Vogel T, Maerzke K A, Iacovella C R, Landau D P, Cummings P T and McCabe C 2013 *J. Chem. Phys.* **139** 54505 1089-7690
[6] Reinhardt A and Frenkel D 2014 *Phys. Rev. Lett.* **112** 238103 0031-9007
[7] Karplus M and McCammon J A 2002 *Nat. Struct. Biol.* **9** 646–652
[8] Liu Y, Prigozhin M B, Schulten K and Gruebele M 2014 *J. Am. Chem. Soc.* **136** 4265–4272
[9] Schelling T C 1971 *J. Math. Sociol.*
[10] Metropolis N, Rosenbluth A W, Rosenbluth M N, Teller A H and Teller E 1953 *J. Chem. Phys.* **21** 1087–1092
[11] Alder B and Wainbright T 1957 *J. Chem. Phys.* **27** 1208
[12] Frenkel D 2013 *Eur. Phys. J. Plus* **128** 10 2190-5444
[13] Perez D, Uberuaga B P, Shim Y, Amar J G and Voter A F 2009 *Chapter 4 Accelerated Molecular Dynamics Methods: Introduction and Recent Developments* vol 5 (Elsevier) ISBN 9780444533593
[14] Sugita Y and Okamoto Y 1999 *Chem. Phys. Lett.* **314** 141–151
[15] Jankowski E and Glotzer S C 2009 *J. Chem. Phys.* **131**
[16] Borovský M, Weigel M, Barash L Y and Žukovič M 2016 *EPJ Web Conf.* **108** 02016 2100-014X
[17] Glaser J, Nguyen T D, Anderson J A, Lui P, Spiga F, Millan J A, Morse D C and Glotzer S C 2014 *Comput. Phys. Commun.* **192** 97–107
[18] Anderson J A, Jankowski E, Grubb T L, Engel M and Glotzer S C 2013 *J. Comput. Phys.* **254** 27–38
[19] Bishop C M 2013 *Pattern Recognition and Machine Learning* vol 53 ISBN 9788578110796 (*Preprint*
[20] Silver D, Huang A, Maddison C J, Guez A, Sifre L, van den Driessche G, Schrittwieser J, Antonoglou I, Panneershelvam V, Lanctot M, Dieleman S, Grewe D, Nham J, Kalchbrenner N, Sutskever I, Lillicrap T, Leach M, Kavukcuoglu K, Graepel T and Hassabis D 2016 *Nature* **529** 484–489 ISSN 0028-0836
[21] Gelly S and Wang Y 2006 URL `http://eprints.pascal-network.org/archive/00002713/`
[22] Long R, Li B, Liu Z and Liu W 2016 *Chem. Eng. J.* **284** 325–332
[23] Ruff K M, Harmon T S and Pappu R V 2015 *J. Chem. Phys.* **143** 243123
[24] Phillips C L and Voth G a 2013 *Soft Matter* **9** 8552 1744-683X
[25] Hachmann J, Olivares-Amaya R, Atahan-Evrenk S, Amador-Bedolla C, Sánchez-Carrera R S, Gold-Parker A, Vogt L, Brockway A M and Aspuru-Guzik A 2011 *J. Phys. Chem. Lett.* **2** 2241–2251
[26] M. Abadi, A. Agarwal, P. Barham, E. Brevdo, Z. Chen, C. Citro, G. Corrado, A. Davis, J. Dean, M. Devin, et al., 2015 TensorFlow: Large-Scale Machine Learning on Heterogeneous Systems (`arXiv:1603.04467v2`)

IOP Conf. Series: Journal of Physics: Conf. Series **921** (2017) 012014 doi:10.1088/1742-6596/921/1/012014

# Interface of topological insulator $Bi_2Se_3$ with $As_2Te_3$

**Niraj Aryal**(1) **and Efstratios Manousakis**(1,2)

(1)Department of Physics and National High Magnetic Field Laboratory, Florida State University, Tallahassee, FL 32306-4350, USA

(2)Department of Physics, University of Athens, Panepistimioupolis, Zografos, 157 84 Athens, Greece

E-mail: manousakis@magnet.fsu.edu

**Abstract.** We study the topological states which appear at the interface of the topological insulator (TI) $Bi_2Se_3$ with $As_2Te_3$ which share similar size lattice constants. A Dirac-like state (DLS) with spin-momentum correlations characteristic to an edge Dirac-like state of a TI, forms at the interface. While our parity analysis shows that one should not expect a topologically protected surface state for $As_2Te_3$, the interface DLS has stronger overlap with the atoms of the $As_2Te_3$ layers, especially slightly away from the zone center. We believe that this interface DLS is the topologically protected $Bi_2Se_3$ edge state which "spills" over to the $As_2Te_3$ layers to optimize its energy.

## 1. Introduction

There has been significant progress towards understanding topological states of matter due to the application of state of the art experimental tools and advanced computational/theoretical techniques. The topological insulator (TI) $Bi_2Se_3$ is an extensively studied system since Zhang *et al.*[1] predicted and Hasan *et al.*[2] verified experimentally its topological signature.

Historically utilization of new phenomena in practical electronic applications is realized through interfaces, such as, a PN junction. In addition, qualitatively different phenomena from bulk can occur and have been observed on interfaces[3, 4]. Furthermore, emergence of such novel phenomena have been theoretically or computationally predicted to occur in interfaces between a TI and a different material[5, 6, 7]. The "interface" of the TI with vacuum, i.e., the TI surface, has been thoroughly studied and rather well understood [8]. While there have been recent studies of the interfacial states formed between a TI and different types of materials, such as, band insulators (e.g., hexagonal Boron-Nitride) [9], semiconductors (e.g., GaAs) [10], graphene[11, 12], etc, our understanding is still nascent to be able to make a device for a practical purpose.

In the present paper, we consider the interface formed between the topological insulator $Bi_2Se_3$ (BS) and $As_2Te_3$ (AT), as illustrated in Fig. 1(a). The $As_2Te_3$ lattice structure is that of the TI family of $Bi_2Se_3$ and $Bi_2Te_3$. Furthermore, the lattice constant of AT is 4.089 Å which is close to that of BS (4.14 Å). Therefore, it looks promising that one can realize such an interface in practice. Unlike BS, in AT the weak spin-orbit coupling due to As atoms is not enough to cause an inversion at the valence and conduction bands near the $\Gamma$ point and, hence, AT is not topological which is confirmed by our parity analysis using the recipe of Ref. [13]. We would like to investigate, as a function of the number of layers, the formation or lack of formation of a topological interface state and to characterize it.

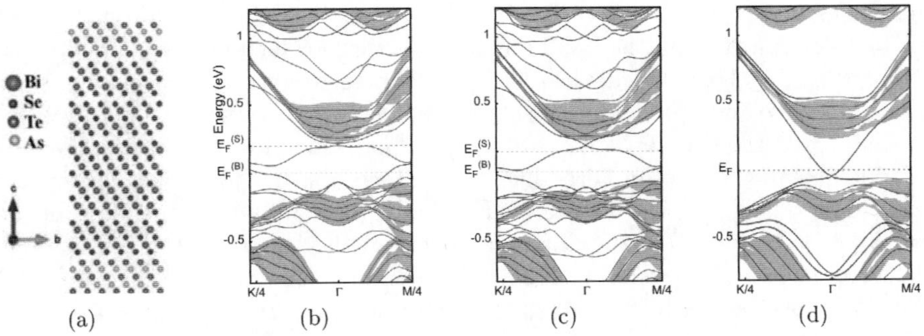


**Figure 1.** (a): The slab which was used in our calculations. (b),(c), and (d): The band structure obtained for various structures. See text for details.

## 2. Computational details

The calculations presented here are carried out using the Quantum Espresso[14] implementation of the density functional theory (DFT) in the GGA framework including spin-orbit coupling (SOC). The Perdew-Burke-Ernzerhof (PBE) exchange correlation functional[15] was used with fully relativistic PAW pseudo-potentials[16].

The $5d$, $6s$ and $6p$ electrons for Bi, the $4s$ and $4p$ electrons for Se, the $4s$ and $4p$ electrons for As and the $4d$, $5s$ and $5p$ electrons for Te were treated as valence electrons in all our calculations.

The plane-wave energy cutoff was taken to be 80 Ry and a $k-$point mesh of $15 \times 15 \times 1$ was used. The slab calculation of the pristine $Bi_2Se_3$ was done using 5 quintuple layers (QLs) whereas the interface calculation was done using 5 and 6 QL of BS with 1 QL of AT on both sides of the slab. A vacuum layer of 17-20 Å was placed to avoid interaction between the periodic images.

We will consider two structures: (a) Structure S1: an AT/BS/AT structure in which we use 5 QLs of $Bi_2Se_3$ and 1 QL of $As_2Te_3$ on each surface of the $Bi_2Se_3$ slab. (b) Structure S2: at AT/BS/AT structure in which we use 6 QLs of $Bi_2Se_3$ and 1 QL of $As_2Te_3$ on each surface of the $Bi_2Se_3$ slab.

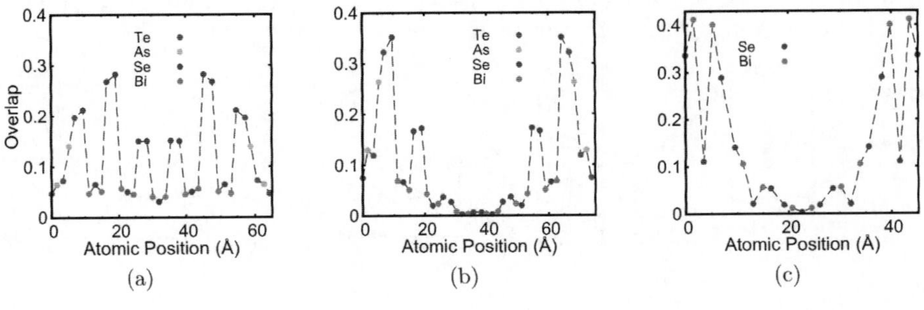


**Figure 2.** Total sum of the projections of all interface Dirac-like states at the Γ point onto the atomic wavefunction of each layer.

## 3. Band structure

Our calculated bands for a slab illustrated in Fig. 1(a), which consists of 1QL-AT/5QL-BS/1QL-AT (which we will refer to as structure S1), are presented in Fig. 1(b). The bands obtained for the same structure as in Fig. 1(a) using 6 BS QLs (which we will refer to as structure S2) are shown in Fig. 1(c). For comparison the band structure of a slab which consists of 5 QLs of pure BS is also presented in Fig. 1(d). In all cases the gray ribbons are the 3D bands of pure crystal BS projected on the a-b plane. First, notice that in the case of the pure slab the bulk ribbons overlap with the states of the slab, except for the Dirac cone which forms in the bulk gap at the Γ point. This topologically protected Dirac-like state which appears for a slab of pure BS has been well-studied previously[8, 1] For the case of Figs. 1(b,c) we shifted all the bands of the slab by a constant energy value equal to the difference between the energy of the bulk ribbons far away from the Fermi level and the energy of the corresponding slab states. Notice that if we do that, the bulk ribbons overlap with the slab states even near the Fermi level and this provides a good guide to the eye for what happens. In Figs. 1(b,c) we denote the slab Fermi level by $E_F^{(S)}$ and the BS bulk Fermi level by $E_F^{(B)}$.

Notice that when we use 6 QLs of BS in the AT/BS/AT interface structure (S2 structure) a Dirac-like surface state forms in the bulk BS gap. As will show later, this state is an interface state and is also characterized by the spin-momentum correlations for an edge state of TI. In the S1 structure (5 BS QLs), there is a small gap separating the top and bottom Dirac cones.

## 4. Nature of the interface bands

Fig. 2 illustrates the degree of localization near the surface of the Dirac-like states (DLS). Notice that in the case of the S2 structure, the DLS are mainly localized near the interface (Fig. 2 (b)) with almost zero contribution from the "bulk" layers, i.e., there are layers away from both interfaces with negligible contribution to the DLS, just like in the pristine BS slab case shown in Fig. 2(c). However, in the case of the S1 structure, the DLS are not just localized near the interface layers (Fig. 2(a)) as their wavefunctions has significant overlap with the atoms of all the layers in the 5 QLs of BS. Namely, the DLS wavefunctions are characterized by a width which is greater than half the size of our BS slab. We believe that this is the reason of the opening of the gap between the upper and lower DLS seen in Fig. 1(b). Therefore, we conclude that a slab with number of QLs greater or equal to 6, is large enough to yield gapless DLS at the interface. In Figs. 3(a,b) we present the character of the bands for the S2 structure. The size of the circles used is proportional to the total projection of the interface state onto the atomic wavefunction of the atoms forming the AT QL (red circles) and the one of the BS QLs which is closer to the interface (blue circles). Notice that the wavefunctions of the DLS have a significant overlap with the atoms of the AT layers. This is also clear from Fig. 2(b) for the Γ point. The contribution of the AT layers becomes greater as we move away from the Γ point.

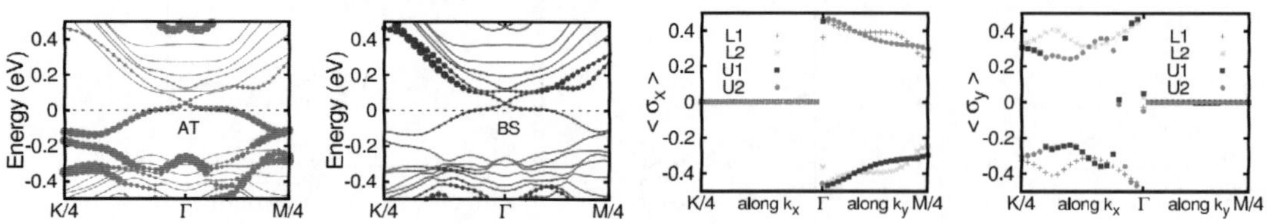


(a) (b) (c) (d)

**Figure 3.** Contribution of the AT layers ((a), red circles) and of the BS layers ((b), blue circles) to DLS. (c) $\langle\sigma_x\rangle$ along $k_y$ and (d) $\langle\sigma_y\rangle$ along $k_x$ for the interface Dirac-like states.

## 5. Spin-momentum correlations of the interface bands

The spin-momentum correlation of a pure BS slab has been calculated by Yazyev *el al.*[17]. In Figs. 3 (c,d) we present the results of our calculations of the spin-momentum correlations for the 1QL AT/6QLs BS/1QL AT system considered in the present paper. We have labeled the curves as X$\nu$ where X=U,L, stands for the upper and lower branches of the Dirac cone and $\nu = 1, 2$, stands for the top and bottom surface of the lab. Notice that the expectation value of $\sigma_x$ along the $k_x$ direction is zero while along the $k_y$ is significant and the upper and lower branch of the Dirac cone are characterized by opposite relative spin orientation and similarly the same branches of the Dirac cone which belong to the top or bottom surface of the slab have opposite spin correlation.

## 6. Conclusions

We have studied the interface of the topological insulator $Bi_2Se_3$ (BS) with a single quintuple layer of $As_2Te_3$ (AT) and we varied the size of the $Bi_2Se_3$ slab. We find that a Dirac-like cone forms inside the bulk BS band gap and its physical location, when the BS slab is large enough, is at the interface. We find that the wavefunction of this interface state has a significant extension inside the AT layer. In fact, we find that as we move away from the $\Gamma$ point, most of the wavefunction of the interface Dirac-like cone lies inside the AT layer. When the number of QLs of BS is 5, the interface Dirac-like state has overlap with every atom of the BS slab, because of the interaction between the two interfaces which are close. In this case the interface Dirac-like cones have a gap. When the number of QLs in the BS slab is 6, we find that the two Dirac-like interface states which form on both interfaces are separated by a spatial gap where their wavefunction is negligible. Therefore, when the thickness of the BS slab is greater or equal to 6 QLs, the Dirac-like interface states are localized at the interfaces. In this case the Dirac cones are gapless. We also calculated the spin-momentum correlations of these interface Dirac-like states and we found that they are the same to those of the topologically protected surface states of the topological insulator.

## 7. Acknowledgments

This work was supported in part by the U.S. National High Magnetic Field Laboratory, which is funded by NSF DMR-1157490 and the State of Florida.

## 8. References

[1] Zhang H, Liu C X, Qi X L, Dai X, Fang Z and Zhang S C 2009 *Nature physics* **5** 438–442
[2] Xia Y, Qian D, Hsieh D, Wray L, Pal A, Lin H, Bansil A, Grauer D, Hor Y, Cava R *et al.* 2009 *Nature Physics* **5** 398–402
[3] Cen C, Thiel S, Hammerl G, Schneider C W, Andersen K E, Hellberg C S, Mannhart J and Levy J 2008 *Nat. Matt.* **7** 278–302
[4] Okamoto S and Millis A J 2004 *Nature* **428** 630
[5] Fu L and Kane C L 2008 *Phys. Rev. Lett.* **100** 096407
[6] Teo J C Y and Kane C L a 2010 *Phys. Rev. B* **82** 115120
[7] Zhang Q, Zhang Z, Zhu Z, Schwingenschlgl U and Cui Y 2012 *ACS Nano* **6** 2345–2352
[8] Hasan M Z and Kane C L 2010 *Rev. Mod. Phys.* **82**(4) 3045–3067
[9] Jin H, Im J, Song J H and Freeman A J 2012 *Phys. Rev. B* **85**(4) 045307
[10] Seixas L, West D, Fazzio A and Zhang S B 2015 *Nature Communications* **6** 7630 EP – article
[11] Zhang J, Triola C and Rossi E 2014 *Phys. Rev. Lett.* **112** 096802
[12] Spataru C D and Léonard F m c 2014 *Phys. Rev. B* **90**(8) 085115
[13] Fu L and Kane C L 2007 *Phys. Rev. B* **76**(4) 045302
[14] Giannozzi P, Baroni S, Bonini N, Calandra M, Car R, Cavazzoni C, Ceresoli D and et al 2009 *Journal of Physics: Condensed Matter* **21** 395502 (19pp)
[15] Perdew J P, Burke K and Ernzerhof M 1996 *Phys. Rev. Lett.* **77**(18) 3865–3868
[16] Blöchl P E 1994 *Phys. Rev. B* **50**(24) 17953–17979
[17] Yazyev O V, Moore J E and Louie S G 2010 *Phys. Rev. Lett.* **105**(26) 266806

IOP Conf. Series: Journal of Physics: Conf. Series **921** (2017) 012015 doi:10.1088/1742-6596/921/1/012015

# Shape matters: The case for Ellipsoids and Ellipsoidal Water[1]

**Andreas F. Tillack[1,2,*] and Bruce H. Robinson[1]**

[1]Dept. of Chemistry, University of Washington, PO 371500, Seattle WA, 98195.

[2]Present address: Center for Computational Sciences, Oak Ridge National Laboratory, One Bethel Valley Road, P.O. Box 2008, MS-6008, Oak Ridge, TN 37831.

[*]Corresponding Author, tillackaf@ornl.gov

**Abstract**. We describe the shape potentials used for the van der Waals interactions between soft-ellipsoids used to coarse-grain molecular moieties in Metropolis Monte-Carlo simulation software. The morphologies resulting from different expressions for these van der Waals interaction potentials are discussed for the case of a prolate spheroid system with a strong dipole at the center. We also show that the calculation of ellipsoids is, at worst, only about fivefold more expensive computationally when compared to a simple Lennard-Jones sphere. Finally, as an application of the ellipsoidal shape we parametrize water from the original SPC water model and observe – just through the difference in shape alone – a significant improvement of the O-O radial distribution function when compared to experimental data.

## 1. Introduction

In the development of coarse-grained force fields based on all atom force fields, the reduction of a set of charges is a reasonably well understood process. Using multipole expansions about centers (or clusters of centers within molecules) one knows how to express many point charges with higher order moments or even fewer partial charges. Additionally, with the use of methods such as CHELPG [1] or RESP [2,3] *ab-initio* quantum mechanical electrostatic potentials can be reduced to a discrete set of point charges (often placed on the atom centers) that well describe most molecules. However, for descriptions of molecular shapes, generally at the all atom level, only spheres are used to represent the van der Waals interactions. Typically, the Lennard-Jones (LJ) type potential is the avatar of van der Waals interactions.

A widely used coarse-grained alternative to a spherical molecular shape is the Gay-Berne formulation of a LJ-like potential energy function implementing spheroidal [4] as well as ellipsoidal shapes [5]. The parameterization of this energy function has proved challenging and there has not been a systematic, bottom-up approach to understanding the parameters needed. We have recently published a method, called the Level-of-Detail method [6], to envision ellipsoids replacing specific units within a molecule, and even the entire molecule. In this work, we described how to calculate the ellipsoid shapes and associated potential parameters directly from an underlying all atom force field such as OPLSAA [7] or GAFF [8]. The results suggest that the ellipsoids and associated parameters that describe the Lennard-Jones energies well reproduce the results of the underlying all atom force field.

This proceedings paper serves as an illustration that ellipsoidal shapes allow for a nuanced representation of the dispersion interactions of molecules. The first part of this paper presents insights

[1]This manuscript has been authored by UT-Battelle, LLC under Contract No. DE-AC05-00OR22725 with the U.S. Department of Energy. The United States Government retains and the publisher, by accepting the article for publication, acknowledges that the United States Government retains a non-exclusive, paid-up, irrevocable, worldwide license to publish or reproduce the published form of this manuscript, or allow others to do so, for United States Government purposes. The Department of Energy will provide public access to these results of federally sponsored research in accordance with the DOE Public Access Plan (http://energy.gov/downloads/doe-public-access-plan).

into how small tweaks to the potential energy function can lead to dramatic changes in the resulting morphology. Secondly, we report on an ellipsoidal water model based on the original SPC [9] model.

## 2. Gay-Berne-like Lennard Jones Potential for Ellipsoids

**Table 1.** Model parameters used in this work

| Model | Semi axes [Å] (LJ width $\sigma_0$ [Å])* | LJ energy [kcal/mol] | Charges; (Locations) [e; Å] | Dipole [Debye] |
|---|---|---|---|---|
| Prolate Spheroid | 12.15x3.14x3.14 (3.0)* | 3.515 | - | (15,0,0) |
| SPC Water [9] | 1.583x1.583x1.583 | 0.155 | -0.82; (0,0,0) | - |
| Water Ellipsoid | 1.973x1.642x1.224 | 0.159 | 0.41; (0.8165,0.5774,0)<br>0.41; (-0.8165,0.5774,0)** | - |

* If "Adjusted Width" LJ potential, eq. 4, is used. ** Charge locations are the same for both models.

Ellipsoids are promising shapes to add to the simulation toolbox [6,10–12]. They enable the description of large, coarse-grained molecular moieties without the need to introduce further corrective terms to the simulation Hamiltonian to account for shape anisotropy [13–15]. Additionally, a physically intuitive understanding of the resulting shapes from simulation parameters is retained. The shape potential used is derived from a generalized Lennard-Jones (LJ) potential similar to the Gay-Berne description [4]:

$$V_{AB}(\vec{r}_{AB},\Omega_{AB}) = 4\epsilon_{AB}\left\{\left(\frac{\sigma_0}{r_{AB}-\delta_{AB}}\right)^{12} - \left(\frac{\sigma_0}{r_{AB}-\delta_{AB}}\right)^{6}\right\} \tag{1}$$

Here, $\vec{r}_{AB}$ is the distance vector (with $r_{AB}$ its magnitude), $\Omega_{AB}$ is the set of relative orientations of the two ellipsoids, $\epsilon_{AB}$ is the potential well depth, $\sigma_0$ determines the width of the potential well, and $\delta_{AB}$ can be used to adjust the zero-crossing of the potential. Note that this reduces to the original LJ potential with $\sigma_0 = const.$ and $\delta_{AB} = 0$. The interaction between two arbitrarily oriented ellipsoids, A and B, is given by $\sigma_{AB}(\vec{r}_{AB},\Omega_{AB})$, as the distance at which the ellipsoids would touch when moved along the distance vector $\vec{r}_{AB}$ from the contact function (CF), $F_{AB}$, described by Perram and Wertheim [16]:

$$\sigma_{AB}(\vec{r}_{AB},\Omega_{AB}) = r_{AB}\cdot F_{AB}(\vec{r}_{AB},\Omega_{AB})^{-\frac{1}{2}} \tag{2}$$

When this expression is used as $\sigma_0$ in conjunction with $\delta_{AB} = 0$ in eq. 1, one obtains a simple, LJ-like potential, called the "Simple Touch" potential [6]:

$$V_{AB}^{CF}(\vec{r}_{AB},\Omega_{AB}) = 4\epsilon_{AB}\left\{\left(\frac{\sigma_{AB}(\vec{r}_{AB},\Omega_{AB})}{r_{AB}}\right)^{12} - \left(\frac{\sigma_{AB}(\vec{r}_{AB},\Omega_{AB})}{r_{AB}}\right)^{6}\right\} = 4\epsilon_{AB}\{F_{AB}^{-6} - F_{AB}^{-3}\} \tag{3}$$

Figure 1a) shows the resulting morphology of identical ellipsoids with a point dipole at their center in a two- and three-dimensional simulation box. The florid, micellar arrangement observed is caused by the wider potential well in the long direction of the ellipsoids. It is a direct consequence of the numerator, $\sigma_{AB}(\vec{r}_{AB},\Omega_{AB})$, in eq. 3 affecting both the zero-crossing as well as the potential well.

Holding the potential well width constant, $\sigma_0 = const.$, and adjusting the zero-crossing such that the ellipsoids cannot overlap, $\delta_{AB} = \sigma_{AB}(\vec{r}_{AB},\Omega_{AB}) - \sigma_0$, leads to the "Adjusted Width" potential:

$$V_{AB}(\vec{r}_{AB},\Omega_{AB}) = 4\epsilon_{AB}\left\{\left(\frac{\sigma_0}{r_{AB}-\sigma_{AB}(\vec{r}_{AB},\Omega_{AB})+\sigma_0}\right)^{12} - \left(\frac{\sigma_0}{r_{AB}-\sigma_{AB}(\vec{r}_{AB},\Omega_{AB})+\sigma_0}\right)^{6}\right\} \tag{4}$$

The resulting morphologies are displayed in Figure 1b). This corrects the micellar behavior observed when using the "Simple Touch" potential in three dimensions.

The remaining, visible ordering behavior in the two-dimensional box is caused by the strong dipole at the center of each ellipsoid as well as by the width of the ellipsoids (in the short direction) and the relative potential well depths. A liquid crystalline arrangement as seen in Figure 1c) would eventually be favored if the circumference of the spheroid were further reduced (increasing dipole-dipole interaction). In the case of Figure 1c) we used the "Interaction Area" orientation dependent potential

IOP Conf. Series: Journal of Physics: Conf. Series **921** (2017) 012015 doi:10.1088/1742-6596/921/1/012015

well depth, $\epsilon_{AB}(\vec{r}_{AB}, \Omega_{AB})$, described in Tillack et al. [6] to favor the side-side interaction over the end-end interaction (by a factor of 3.87). This approach takes into account the dispersion interaction's proportionality to the surface areas of the interacting ellipsoids.

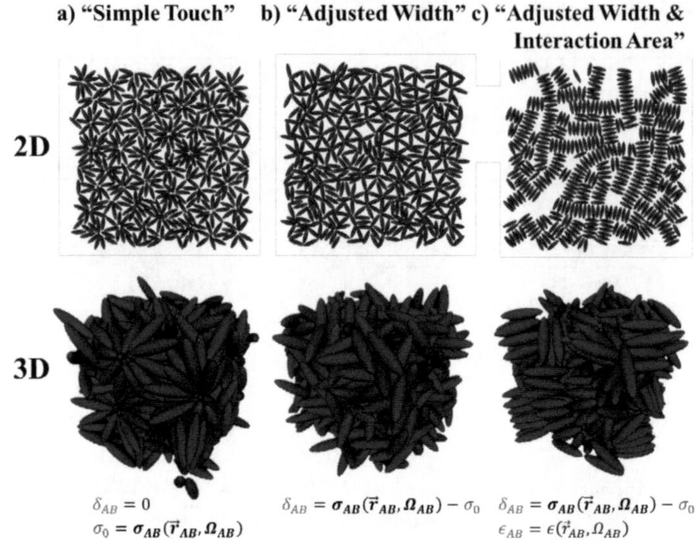


**Figure 1.** Effect of choice of Lennard-Jones Potential on resulting arrangement of ellipsoid-dipole system in two-dimensional (2D, top) and three-dimensional (3D, bottom; adapted with permission from [6]. Copyright 2016 American Chemical Society) simulation box. System consists of 432 prolate ellipsoids (see Table 1) in the the isothermal isobaric (NPT) ensemble under 1 atm at 400 K with periodic boundary conditions (PBCs). The 2D system was obtained by using a modified soft wall restoring potential [17]; walls separated by $L = \pm 2 \cdot 3.14$ Å (matching the spheroid's short axis); a potential well depth of $6.6 \cdot 10^{-3} \frac{kcal}{mol}$; and PBCs in the y and z-direction.

An often thought downside to using ellipsoids and particularly when using Perram and Wertheim's contact function [16] is that the numerical implementation is done in an iterative fashion with no clearly specified end-condition, which implies an indeterminate computational cost. This naturally makes some feel queasy because, theoretically, this could mean one ends up with an infinite loop in certain cases. However, if one fixes the precision of the answer (in our case to a variance of $10^{-6}$ in the scalar parameter $\lambda$ of the contact function [16]) not only does the calculation typically converge within 5-6 loops, but also fewer iterations are needed the farther ellipsoids are from each other. The same lowered computational cost also applies for ellipsoids with smaller aspect ratios.

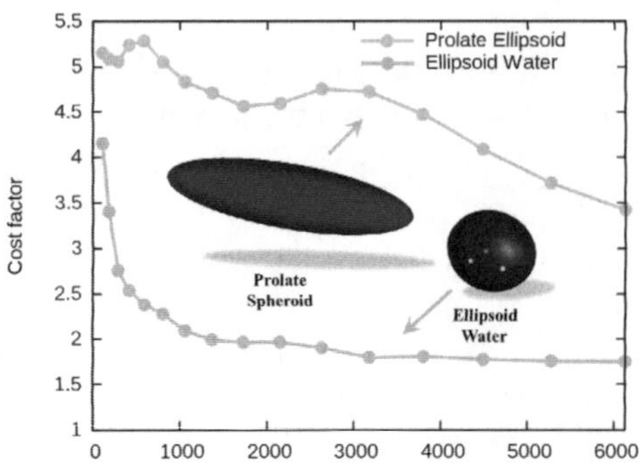


**Figure 2.** Cost factor of ellipsoids, from Table 1, is the cost (i.e. time) of computation relative to an equally sized system of a Lennard-Jones sphere. All calculations were run in the canonical (NVT) ensemble with potential cut off lengths fixed to half the box length.

Figure 2 shows the relative cost of using the ellipsoids from Table 1 with equation 2 but no electrostatics. For an ellipsoid of small aspect ratio (ellipsoidal water) the cost factor for increasing system sizes, $N$, decreases towards an asymptotic value of around 1.75. For the prolate ellipsoids, this drop in calculation cost is still ongoing for the largest system of 6000 ellipsoids. Overall, the maximum cost factor is about five, consistent with previous work [6], compared to the calculation of the spherical Lennard-Jones potential. Because the pairwise evaluation scales with the square of the number of entities involved, then when three or more Lennard-Jones spheres are replaced by one ellipsoid, the ellipsoid will be more computationally efficient ($\sqrt{5.5} < 3$, where 5.5 is the maximum observed ellipsoid cost factor). Ellipsoidal potentials may be adapted to an MD software package, because the contact function is analytic and could be used to evaluate forces and torques for both equations 3 and 4 [18].

## 3. Water: The Importance of Shape

To illustrate the utility of using an ellipsoidal shape, let us consider a simple, ellipsoidal water model based on the original SPC model [9]. By using the gyration tensor approach outlined in [6] an ellipsoid can be obtained from the SPC water model with oxygen, O, at the center of the ellipsoid.

With a Hamiltonian consisting of equation 3 for the LJ-like pair potential as well as a Coulomb interaction potential and a Barker-Watts type self-consistent reaction field [10,12,19] the resulting radial distribution function between oxygen centers for both the original SPC [9] and the ellipsoid water model in comparison to current experimental data from [20] is shown in Figure 3. It can be observed that while SPC generally matches the secondary peak locations well, it does overshoot the main peak by about 12% and generally shows slightly broader peaks than seen experimentally. The ellipsoid model improves the radial distribution function significantly in those regards.

Both models, using the NPT ensemble, give similar densities of $(0.98 \pm 0.01)\ g/cc$. Interestingly, while the SPC model gave an overall dielectric of $60 \pm 20$ the ellipsoid model returned a dielectric of $23 \pm 5$. This dramatic change of dielectric constants between the two models can be exclusively attributed to the difference in shape, spherical versus ellipsoidal. The ellipsoid breaks the high symmetry offered by the sphere used in the SPC water model. Combined with the strong dipole component of the interior charge distribution this leads the SPC water model (and by extension any water models using a spherical shape) to reach a larger dielectric constant than would be possible with a lower symmetry shape.

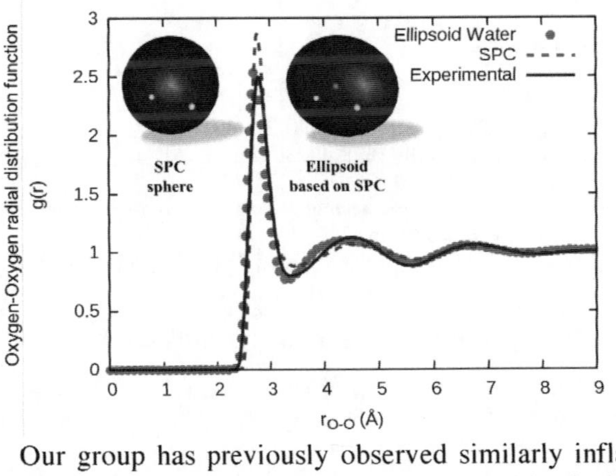


**Figure 3.** Shape effect demonstrated using O-O radial distribution function, g(r), of water with spherical SPC water model (dashed red line), ellipsoidal model (green dots) based on SPC in comparison to experimental data[20] (solid blue line). Simulations consist of 216 water molecules (see Table 1) in the NPT ensemble under 1 atm at 298 K with periodic boundary conditions.

Our group has previously observed similarly inflated dielectric constants for a Stockmeyer fluid (dipole in a sphere) when compared to an identical dipole inside of a prolate spheroid [10]. More importantly, prolate spheroids, with an aspect ratio larger than 1.3, agreed well with the Onsager value of the dielectric (quite close to the experimental dielectric for most organic solvents). Furthermore, using a simple dipolar expansions of charges and proper ellipsoids representing ethylene carbonate, we reported that analogous calculations obtained dielectric constants of 90, in agreement with experiment, using only a single or a two-ellipsoid coarse-grained model [6,14]. Therefore, it should be feasible to obtain the experimental dielectric using the modified water shape, and position the charges to best describe the electrostatic potential of water, as recently suggested [21].

## 4. Conclusions

We have shown dramatic differences in the morphology of prolate spheroids with large dipoles, having a dipolar density on the order of water (5.42 for the prolate spheroid at 400 K vs 5.83 for SPC water at 298 K). The different morphologies shown for the prolate spheroids are caused by their large aspect ratio. Water however, has a much smaller aspect ratio but subtleties in close interactions can manifest in small changes in g(r). In this case the simple substitution of an ellipsoid for the sphere in water definitely improved the SPC model of water with respect to the experimental g(r). No further modifications of the model were performed.

The modification of the molecular shape to generate an ellipsoid for the Lennard-Jones interactions is based on an all atom transferable force field. Using well developed rules, one can automatically replace any simple set of atoms (within a molecule) with an ellipsoid. In the several cases we have tested [6] the agreement of the g(r) of the ellipsoidal coarse-grained representation with the underlying all atom g(r) has been excellent. We point out with water, that the ellipsoid we developed was consistent with the symmetry of the water molecule. The use of a sphere produces a system with a higher symmetry in its shape. The interaction of the point charges in the SPC Hamiltonian does respect the molecular symmetry of the molecule, but the same cannot be said for the intermolecular van der Waals energies. Replacing a sphere with an ellipsoid restores the underlying molecular dispersion symmetry of the water.

## Acknowledgements

We gratefully acknowledge the financial support of the National Science Foundation (DMR-1303080); and the Air Force Office of Scientific Research (FA9550-10-1-0558 and FA9550-15-1-0319). Any opinions, findings, and conclusions or recommendations expressed in this material are those of the author(s) and do not necessarily reflect the views of the National Science Foundation. Andreas F. Tillack has been partially supported by the U.S. Department of Energy, Office of Science, Office of Advanced Scientific Computing Research, Oak Ridge Leadership Computing facility under contract number DE-AC05-00OR22725.

## References

[1] Breneman C M and Wiberg K B 1990 *J. Comput. Chem.* **11** 361–73
[2] Bayly C I, Cieplak P, Cornell W and Kollman P A 1993 *J. Phys. Chem.* **97** 10269–80
[3] Cornell W D, Cieplak P, Bayly C I and Kollmann P A 1993 *J. Am. Chem. Soc.* **115** 9620–31
[4] Gay J G and Berne B J 1981 *J. Chem. Phys.* **74** 3316
[5] Berardi R, Fava C and Zannoni C 1995 *Chem. Phys. Lett.* **236** 462–8
[6] Tillack A F, Johnson L E, Eichinger B E and Robinson B H 2016 *J. Chem. Theory Comput.* **12** 4362–74
[7] Jorgensen W L, Maxwell D S and Tirado-Rives J 1996 *J. Am. Chem. Soc.* **118** 11225–11236
[8] Wang J, Wolf R M, Caldwell J W, Kollman P A and Case D A 2004 *J. Comput. Chem.* **25** 1157–74
[9] Berendsen H J C, Postma J P M, Gunsteren W F van and Hermans J 1981 *Intermolecular Forces* The Jerusalem Symposia on Quantum Chemistry and Biochemistry ed B Pullman (Springer Netherlands) pp 331–42
[10] Johnson L E, Barnes R, Draxler T W, Eichinger B E and Robinson B H 2010 *J. Phys. Chem. B* **114** 8431–40
[11] Johnson L E 2012 (Seattle: University of Washington)
[12] Tillack A F 2015 (Seattle: University of Washington)
[13] Benight S J, Johnson L E, Barnes R, Olbricht B C, Bale D H, Reid P J, Eichinger B E, Dalton L R, Sullivan P A and Robinson B H 2010 *J. Phys. Chem. B* **114** 11949–56
[14] Tillack A F, Johnson L E, Rawal M, Dalton L R and Robinson B H 2014 *Symposium II/JJ/KK – Materials, Processes and Devices for Nanophotonics, Nonlinear Optics and Resonant Optics* MRS Online Proceedings Library Archive vol 1698p mrss14-1698-jj08-05 (6 pages)
[15] Tillack A F and Robinson B H 2016 *JOSA B* **33** E121–9
[16] Perram J W and Wertheim M S 1985 *J. Comput. Phys.* **58** 409–416
[17] Heni W, Haffner C, Elder D L, Tillack A F, Fedoryshyn Y, Cottier R, Salamin Y, Hoessbacher C, Koch U, Cheng B, Robinson B, Dalton L R and Leuthold J 2017 *Opt. Express* **25** 2627–53
[18] Paramonov L and Yaliraki S N 2005 *J. Chem. Phys.* **123** 194111
[19] Allen M P and Tildesley D J 1989 (Oxford, England; Oxford University Press)
[20] Soper A K 2013 *Int. Sch. Res. Not.* **2013** e279463
[21] Izadi S, Anandakrishnan R and Onufriev A V 2014 *J. Phys. Chem. Lett.* **5** 3863–71

# Yaldram-Khan model with CO and N desorption

**Jonás Díaz**[1] **and Gloria María Buendía**[1]

[1]Department of Physics, Universidad Simón Bolívar, Apartado 89000, Caracas 1080, Venezuela

E-mail: buendia@usb.ve

**Abstract.** We present kinetic Monte Carlo simulations of a modified Yaldram-Khan (YK) model for the catalytic reduction of NO on a surface. The model includes the individual desorption of CO molecules and N atoms, associated with temperature effects. We find that the combined desorption of these species produces a reactive phase that is absent in the original YK model on a square lattice. In this reactive phase, N checkerboard structures are reduced in size and reactive zones between them can be established.

## 1. Introduction

The catalytic reduction of NO has received a great deal of attention because of the key role it plays in the control of vehicle exhaust emissions [1, 2], and the variety of non-equilibrium phenomena it can exhibit [3, 4, 5, 6]. A simple lattice-gas model, based on the Langmuir-Hinshelwood mechanism, was proposed by Yaldram and Khan (YK) [7] to study this reaction. In this model, inspired by the Ziff-Gulari-Barshad model for the catalytic oxidation of CO [8], CO and NO molecules in a gas phase, can be adsorbed on a catalytic surface and once there react. The adsorption probability depends on the partial pressure of the compound in the gas mixture, $y$ for CO and $1 - y$ for NO. CO molecules are adsorbed on single sites, while NO molecules are dissociatively adsorbed on nearest-neighbor (nn) sites. When CO and O or a pair of N atoms are located on nn sites, they react, liberating $CO_2$ or $N_2$, respectively, and leaving behind two empty sites. On a square lattice, the formation of checkerboard structures of N atoms prevents the existence of a reactive state [9]. Several authors have tried to eliminate this "checkerboarding" process on a square lattice, by modifying the original reaction by the inclusion of CO desorption or adding an Eley-Rideal mechanism [10, 11, 12, 13, 14, 15].

Other approaches to the NO reduction are provided by more elaborate lattice-gas models that explicitly take into account energetic effects [16, 17, 18]. Still, we consider that there is much to learn from a simple model such as the YK that can be easily modified to include many relevant processes. In this paper we study the individual desorption of CO molecules and N atoms in the YK model on a square lattice. We are not aware of previous studies that take into account the desorption of N atoms in the CO+NO reaction.

## 2. Model and Simulations

The model is simulated on a square lattice that represents the catalytic surface. A Monte Carlo simulation generates a sequence of trials: CO or NO adsorption, or CO or N desorption. A schematic representation of the process is given by the following equations:

$$
\begin{aligned}
CO(g) + S &\rightarrow CO(a) \\
NO + 2S &\rightarrow N(a) + O(a) \\
CO(a) + O(a) &\rightarrow CO_2(g) + 2S \\
N(a) + N(a) &\rightarrow N_2(g) + 2S \\
CO(a) &\rightarrow CO(g) + S \\
N(a) &\rightarrow N(g) + S
\end{aligned}
\qquad (1)
$$

Here S represents an empty site on the surface, g means gas phase and a means adsorbed. The first four steps correspond to the Langmuir-Hinshelwood mechanism that defines the YK model. The last two correspond to the desorption processes that we are including. Coverage is defined as the fraction of sites on the surface occupied by an adsorbate. We calculate the CO, O and N coverages, and the rate of production of $CO_2$ and $N_2$. $\theta_X$ stands for the coverage of species X.

## 3. Results

We first fix the value of $k_N$ and vary $k_{CO}$. In Fig. 1 and Fig. 2 we present the coverages and reaction rates, respectively, for $k_N = 0.1$ and different $k_{CO}$ values. For small values of $y$ the behavior of the system is practically independent of $k_{CO}$. In this low $y$ region, the coverage of CO is very low, so its desorption is not relevant. For $k_{CO} = 0$, there is a discontinuous transition from a reactive state to a CO poisoned one. As $k_{CO}$ increases, the transition becomes smoother and there is no CO poisoning. For large enough values of $k_{CO}$, the transition seems to be continuous, $\theta_{CO}$ increases almost linearly with $y$ while $\theta_O$ correspondingly decreases in a linear way. $\theta_N$ reaches a maximum and then decreases smoothly toward zero. $\theta_{CO}$ decreases with $k_{CO}$ in favor of $\theta_N$ and $\theta_O$ in such a way that now a reactive state is possible.

CO desorption clearly favors the reactivity of the system, increasing with $k_{CO}$ as seen in Fig. 2. This is expected since now the lattice cannot be poisoned with CO. There is an interval of $y$ within the reactive window for which the CO coverage is almost zero, indicating that in that region most of the CO atoms once adsorbed react immediately with O atoms. Analogously, there is an interval, for higher values of $y$, for which the O coverage is almost zero. In this region the O atoms are the ones that react immediately with CO molecules. In this fashion, particularly for high values of $y$, O atoms are depleted from the surface, while some N atoms still remain.

When $k_{CO}$ increases, the production rates increase as well, in spite of the fact that the N structures tend to grow in size. The production takes place in the regions between such structures. Within these reactive regions, a larger CO desorption rate favors the reactivity because there are more empty sites available where NO molecules can adsorb. At the same time, this favors the creation of N structures but the inclusion of $k_N$ keeps them from covering the lattice and establishing a non-reactive state.

Next, we fix the value of $k_{CO}$ and vary $k_N$. In Fig. 3 and Fig. 4 we present the coverages and reaction rates, respectively, for $k_{CO} = 0.3$ and different $k_N$ values. $\theta_{CO}$ increases with $k_N$ mainly to the detriment of $\theta_N$. $\theta_O$ remains rather unaltered. In this case a reactive state also appears, and the reactive window increases its area as $k_N$ increases. These observations suggest that desorbing N atoms, adjacent to O atoms, leave individual empty sites where CO molecules can adsorb and react. For high values of $y$ this leads to a lattice almost covered with CO. In contrast to the case with $k_N$ fixed, the positions of the maxima of production are almost independent of $k_N$.

$N_2$ production increases initially with $k_N$ but, unlike $CO_2$ production, it reaches a saturation value. This happens rather quickly and the reaction rates of $N_2$ are almost the same from

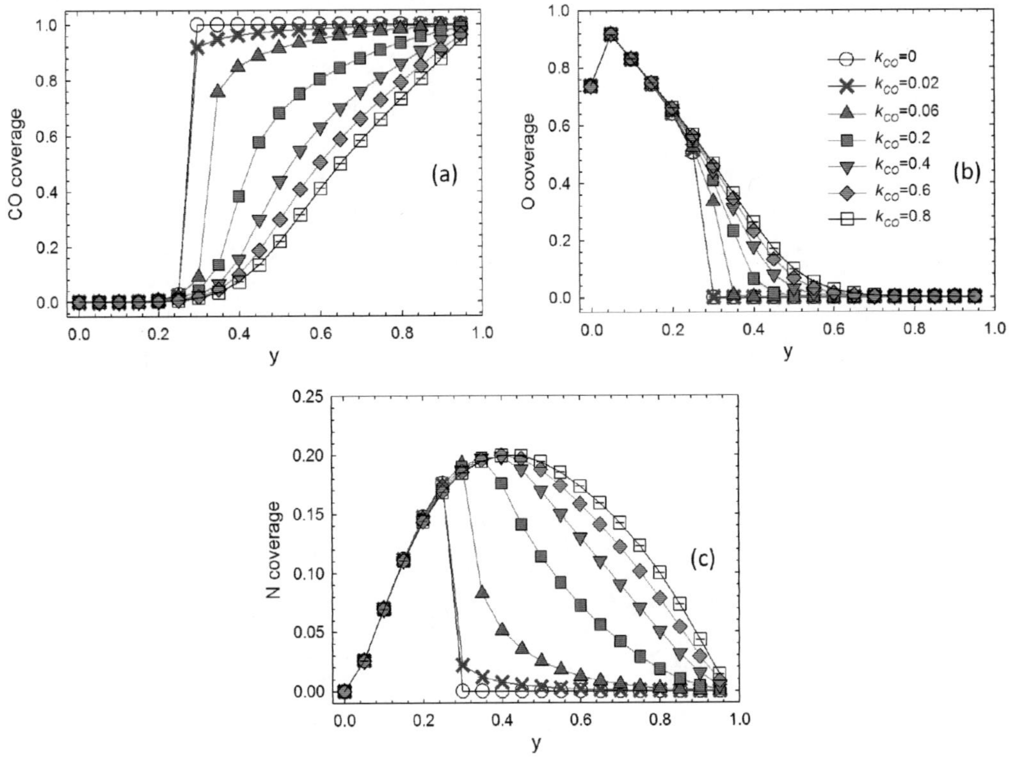


**Figure 1.** Coverages vs $y$ when there is CO and N desorption with $k_N = 0.1$ fixed. (a) CO (b) O and (c) N. The values of $k_{CO}$ are indicated in (b).

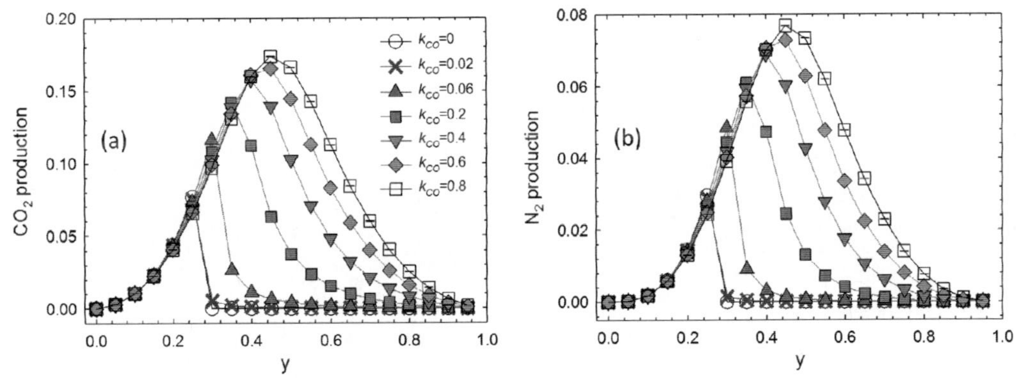


**Figure 2.** Reaction rates vs $y$ when there is CO and N desorption with $k_N = 0.1$ fixed. The values of $k_{CO}$ are indicated in (a).

$k_N = 0.05$ onward. This happens because the number of N atoms capable of reacting on the lattice is reduced by N desorption. The curves with $k_N = 0$ are not very smooth, because the system with only $k_{CO}$ presents large fluctuations associated with the growth of $N$ structures that may or may not cover the entire lattice [19].

Snapshots of the lattice show, as expected, that the formation of N structures is hampered by the inclusion of N desorption. For low $k_N$ values, some N structures remain and the production

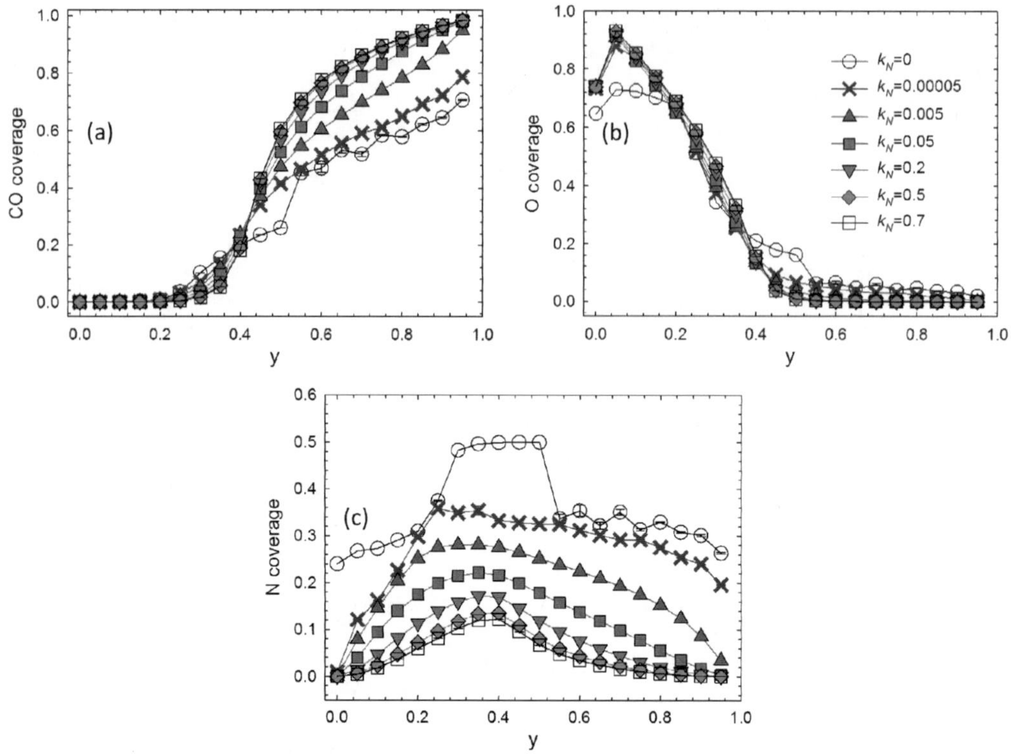


**Figure 3.** Coverages vs $y$ when there is CO and N desorption with $k_{CO} = 0.3$ fixed. (a) CO (b) O and (c) N. The values of $k_N$ are indicated in (b).

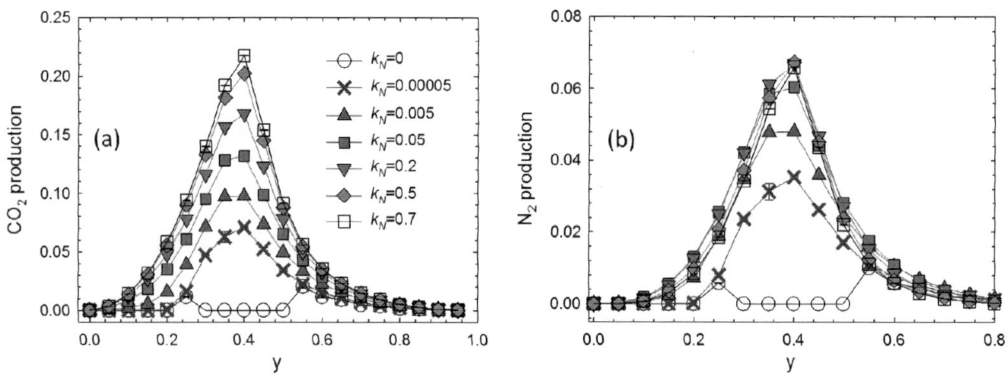


**Figure 4.** Reaction rates vs $y$ when there is CO and N desorption with $k_{CO} = 0.3$ fixed. The values of $k_N$ are indicated in (a).

takes place in the regions between them. For high $k_N$ values, the N structures disappear almost completely, and the production of $CO_2$ and $N_2$ occurs uniformly over the lattice. When N desorption is present, the production of $N_2$ depends on two competing effects: the reduction of the N structures that favors the reactivity of the lattice, and the reduction of the number of N atoms capable of reacting that minimizes the $N_2$ reaction rate.

## 4. Conclusions

In this work we studied the catalytic reduction of NO on a surface by kinetic Monte Carlo simulations of a modified YK model that includes CO and NO desorption rates, $k_{CO}$ and $k_N$, respectively. We found that, contrary to what happens in the standard YK model on a square lattice, our modified model presents a reactive window where $CO_2$ and $N_2$ are produced. For a fixed value of $k_N$, an increment of $k_{CO}$ impedes the formation of a CO poisoned state and favors the formation of N structures. The reactivity is enhanced because larger $k_{CO}$ values benefit the reactivity since more empty sites are available where NO molecules can adsorb, the N desorption stops the N structures from taking over the lattice, and reactive zones between such structures can be established. For a fixed value of $k_{CO}$, both the production of $CO_2$ and $N_2$ are favored by increasing $k_N$ but not in the same proportion. This is in contrast to the case with $k_N$ fixed, where the increase of $k_{CO}$ favors both reaction rates equally.

## References

[1] Taylor K C 1993 *Catal. Rev. Sci. Eng.* **35** 457
[2] Liu G and Gao P 2011 *Catal. Sci. Technol.* **1** 552
[3] Marro J and Dickman R 1999 *Nonequilibrium phase transitions in lattice models* (New York: Cambridge University Press).
[4] Christmann K 1991 *Introduction to Surface Physical Chemistry* (Darmstadt: Steinkopff Verlag).
[5] Bond G C 1987 *Catalysis: Principles and Applications* (Oxford: Clarendon Press).
[6] Imbihl R and Ertl G 1995 *Chem. Rev.* **95** 697
[7] Yaldram K and Khan M A 1991 *J. Catal.* **131** 369
[8] Ziff R M, Gulari E and Barshad Y 1986 *Phys. Rev. Lett.* **56** 2553
[9] Brosilow B J and Ziff R M 1992 *J. Catal.* **136** 275
[10] Meng B, Weinberg W H and Evans J W 1993 *Phys. Rev. E* **48** 3577
[11] Meng B, Weinberg W H and Evans J W 1994 *J. Chem. Phys.* **101** 3234
[12] Kortlüke O and von Niessen W 1996 *J. Chem. Phys.* **105** 4764
[13] Khan M A, Yaldram K, Khalil G K and Khan K M 1994 *Phys. Rev. E* **50** 2156
[14] Khan M A and Ahmad W 2002 *J. Phys. A: Math. Gen.* **35** 2713
[15] Bustos V, Uñac R, Zaera F and Zgrablich G 2003 *J. Chem. Phys.* **118** 9372
[16] Luque J J, Gómez A and Córdoba A 2004 *Physica A* **31** 505
[17] Cortés J and Valencia E 2008 *Bull. Chem. Soc. Jpn.* **81** 1267
[18] Mei D H, Du J C and Neurock M 2010 *Ind. Eng. Chem. Res.* **49** 10364
[19] Díaz J J, and Buendía G M 2017 Submitted

# Population annealing: Massively parallel simulations in statistical physics

**Martin Weigel[1], Lev Yu. Barash[2,3], Michal Borovský[4], Wolfhard Janke[5], and Lev N. Shchur[2,3,6]**

[1] Applied Mathematics Research Centre, Coventry University, Coventry, CV1 5FB, United Kingdom
[2] Landau Institute for Theoretical Physics, 142432 Chernogolovka, Russia
[3] Science Center in Chernogolovka, 142432 Chernogolovka, Russia
[4] P.J. Šafárik University, Park Angelinum 9, 040 01 Košice, Slovak Republic
[5] Institut für Theoretische Physik, Universität Leipzig, Postfach 100920, 04009 Leipzig, Germany
[6] National Research University Higher School of Economics, 101000 Moscow, Russia

E-mail: martin.weigel@complexity-coventry.org

**Abstract.**
The canonical technique for Monte Carlo simulations in statistical physics is importance sampling via a suitably constructed Markov chain. While such approaches are quite successful, they are not particularly well suited for parallelization as the chain dynamics is sequential, and if replicated chains are used to increase statistics each of them relaxes into equilibrium with an intrinsic time constant that cannot be reduced by parallel work. Population annealing is a sequential Monte Carlo method that simulates an ensemble of system replica under a cooling protocol. The population element makes it naturally well suited for massively parallel simulations, and bias can be systematically reduced by increasing the population size. We present an implementation of population annealing on graphics processing units and discuss its behavior for different systems undergoing continuous and first-order phase transitions.

## 1. Introduction

Population annealing (PA) was first suggested in 2001 by Iba [1] and later on discussed in more detail by Hukushima and Iba [2] as a method to tackle potentially difficult sampling problems, but with no particular view to a parallel implementation. It belongs to a family of population methods with examples in quantum Monte Carlo ("diffusion Monte Carlo" [3]), statistical inference ("particle filter" [4]) and statistical physics ("go with the winners" [5]), among others. PA considers an ensemble of configurations set up in equilibrium at high temperature. These are then propagated down to low temperatures by a combination of local updates and a resampling or population control step that clones or prunes configurations according to their relative statistical weight. PA is hence not a Markov chain method but belongs to the different realm of sequential Monte Carlo techniques [6] that are quite commonly used in statistics, but less well known in physics.

It is mostly due to limitations in power dissipation that the continuous increase in clock frequencies of commodity processors has come to an end more than 10 years ago. The computational power of a typical PC or cluster node has continued to increase, however, but now fueled by a multiplication of the cores available for computation, i.e., through an increase in parallelism [7]. As a consequence, high-performance computing clusters can now come with millions of cores, and an end to this development towards ever more parallel computational resources is not in sight. The search for algorithms that perform well in such environments is hence one of the main tasks of computational science today [8]. In Monte Carlo, Markov chain methods that are so successful in serial and moderately parallel setups are not a perfect fit for massively parallel computational resources: parallelization there is either through domain decomposition with the obvious limitations in strong scaling as the domains become too small, or through the simulation of independent or loosely coupled parallel walkers to increase statistics [9–11]. In the latter approach the fraction of time spent outside of the equilibration phase shrinks with the number of parallel threads, leading to asymptotically vanishing efficiency. In PA, on the other hand, the quality of approximation improves with the population size $R$, with systematic deviations (bias) asymptotically decaying proportional to $1/R$ [12, 13]. As a consequence, this approach and its variations hold great promise for becoming standard simulation techniques for the age of massively parallel computing.

The study of systems undergoing phase transitions and, in particular, those with complex free-energy landscapes has been one of the main application areas and one of the main driving forces of algorithm development within the realm of computer simulations in statistical physics [14]. The effect of critical slowing down observed in the vicinity of continuous phase transitions led to the development of cluster updates [15]. Phase coexistence and the suppression of the region of mixed states connecting the pure phases for first-order transitions became amenable to quantitative studies through the multicanonical method and related approaches [16, 17]. On the other hand, frustrated systems with many metastable states can be simulated (more) efficiently through the use of parallel tempering [18] and similar techniques. This latter case of systems with complex free-energy landscapes is where PA has been recently used with some success, namely for the simulation of spin glasses [13, 19, 20]. A more general understanding of the performance of PA for simulations of systems undergoing continuous and discontinuous phase transitions is lacking to date.

In the following we show how the highly parallel resources provided by graphics processing units (GPUs) can be used to achieve an efficient implementation of PA for the case of the Potts model. We then study the behavior of the algorithm for simulations of the square-lattice Potts model, distinguishing the regimes where it undergoes second-order and where it undergoes first-order phase transitions, respectively.

## 2. Population annealing

Population annealing as a weighted sequential algorithm was first introduced in Refs. [1, 2]. The variant which we discuss here follows the scheme proposed more recently by Machta [21]. The algorithm is defined by the following steps:

(i) Set up an equilibrium ensemble of $R_0 = R$ independent replicas of the system at inverse temperature $\beta_0$. Often one chooses $\beta_0 = 0$, where this can be easily achieved.

(ii) To create an approximately equilibrated population at $\beta_i > \beta_{i-1}$, resample configurations $j = 1, \ldots, R_{i-1}$ with their relative Boltzmann weight $\tau_i(E_j) = \exp[-(\beta_i - \beta_{i-1})E_j]/Q_i$, where

$$Q_i \equiv Q(\beta_{i-1}, \beta_i) = \frac{1}{R_{i-1}} \sum_{j=1}^{R_{i-1}} \exp[-(\beta_i - \beta_{i-1})E_j]. \quad (1)$$

(iii) Update each replica by $\theta$ sweeps of a Markov chain Monte Carlo (MCMC) algorithm at inverse temperature $\beta_i$.

(iv) Calculate estimates for observable quantities $\mathcal{O}$ as population averages $\sum_{j=1}^{R_i} \mathcal{O}_j / R_i$.

(v) If the the target temperature $\beta_{\rm f}$ has not been reached, goto step (ii).

If one starts with an equilibrium sample, for instance by using a set of random configurations at infinite temperature, one arrives at a valid algorithm even without the additional rounds of sampling in step (iii). As in each temperature step some of the properly normalized resampling factors will be smaller than unity, however, this would lead to an exponential decrease in the number of distinct replicas and hence a vanishing efficiency in representing the target distribution. A combination with MCMC updates is therefore crucial for the approach, which hence becomes a hybrid of sequential and Markov chain methods.

The resampling step is very similar to what happens in histogram reweighting [22], i.e., taking account of the non-unit weight of an equilibrium configuration at inverse temperature $\beta$ for the ensemble at temperature $\beta'$. As these weights are immediately translated into a resampling of the population, in the present implementation individual configurations all carry the same weight [2]. The expected number of copies of each replica at $\beta_{i-1}$ to be included in the population at $\beta_i$ should be $\hat{\tau}_i(E_j) = (R/R_{i-1})\tau_i(E_j)$. Other than fixing the expectation values, there is some freedom of choice for the actual probability distribution of the number $r_i^j$ of copies of replica $j$ in the population at inverse temperature $\beta_i$. The two most common cases are a Poisson distribution [21],

$$r_i^j \sim \mathrm{Pois}[\hat{\tau}_i(E_j)], \tag{2}$$

or a fixed integer part plus a Bernoulli trial corresponding to the fractional contribution [23, 24],

$$r_i^j = \begin{cases} \lfloor \hat{\tau}_i(E_j) \rfloor & \text{if } r > \hat{\tau}_i(E_j) - \lfloor \hat{\tau}_i(E_j) \rfloor \\ \lfloor \hat{\tau}_i(E_j) \rfloor + 1 & \text{otherwise} \end{cases} . \tag{3}$$

Here, $r$ is a random number uniformly distributed in $[0, 1)$ drawn for each replica of the population at $\beta_{i-1}$, and $\lfloor x \rfloor$ denotes the largest integer not exceeding $x$.

The temperature protocol is not formally restricted in the algorithm as outlined above. The simplest choice corresponds to a constant inverse temperature step $\Delta\beta$, $\beta_i = \beta_{i-1} + \Delta\beta$. More advanced schemes are possible, however, such as an adaptive choice of temperature steps guaranteeing sufficient overlap of neighboring energy histograms [25]. A particular feature of the PA algorithm is that it provides a natural estimator of the free energy through a combination of the resampling factors $Q_k$ [21],

$$-\beta_i F(\beta_i) = \ln Z_{\beta_0} + \sum_{k=1}^{i} \ln Q_k. \tag{4}$$

Here $Z_{\beta_0}$ denotes the partition function at the initial temperature point $\beta_0$ which is required in order to get absolute free energies. For $\beta_0 = 0$ it is easily found as it corresponds to the case of the non-interacting system.

## 3. Implementation on GPU

As a sufficiently versatile example for the study of phase transitions of first and second order, we considered the Potts model on the square lattice with Hamiltonian [26]

$$\mathcal{H} = -J \sum_{\langle i,j \rangle} \delta_{s_i, s_j}, \tag{5}$$

where $s_i \in \{1, \ldots, q\}$ and $J > 0$ corresponds to a ferromagnetic coupling (we choose units such that $J = 1$). As indicated in the sum in Eq. (5) we restrict interactions to nearest neighbors only, and we assume periodic boundary conditions. The model has a phase transition at the

IOP Conf. Series: Journal of Physics: Conf. Series **921** (2017) 012017 doi:10.1088/1742-6596/921/1/012017

**Table 1.** Optimal performance of the CPU and GPU codes for PA for the $q = 2$ model in units of the time $t_{\rm SF}$ per spin flip. As the best performance is found for large $\theta$, we quote here data for $\theta = 500$ and use $R = 50\,000$. The speedups for the single-spin coding (SSC) and multi-spin coding (MSC) GPU programs are relative to the CPU results. GPU performance data are for the Tesla K80 device. The CPU code was benchmarked on an Intel Xeon E5-2683 v4 CPU running at 2.1 GHz.

| | CPU | GPU | | | |
|---|---|---|---|---|---|
| | | SSC | | MSC | |
| $L$ | $t_{\rm SF}$ [ns] | $t_{\rm SF}$ [ns] | speedup | $t_{\rm SF}$ [ns] | speedup |
| 16 | 23.1 | 0.094 | 246 | 0.0096 | 2406 |
| 32 | 22.9 | 0.092 | 249 | 0.0095 | 2410 |
| 64 | 22.6 | 0.092 | 246 | 0.0098 | 2306 |
| 128 | 22.6 | 0.097 | 233 | 0.0098 | 2306 |
| 256 | 22.5 | 0.098 | 230 | 0.0099 | 2273 |

self-dual point $\beta_t = \ln(1+\sqrt{q})$ [27, 28] which is of second order for $q \leq 4$ and of first order for $q > 4$ [29, 30]. The Ising model is equivalent to the special case $q = 2$ as becomes clear on making the identification $\sigma_i = 2s_i - 3$ between Ising spins $\sigma_i$ and the Potts spins $s_i$.

The PA algorithm outlined above consists of the three repeating steps (ii) population resampling, (iii) MCMC updates, and (iv) measurement cycle. In typical applications, one mostly chooses $\theta \gtrsim 10$, and as a consequence the MCMC updates typically take more than 80% of the time. Hence their efficient implementation is the most significant element [31]. We chose the Metropolis algorithm to update the spins [14]. This part of the simulation is trivially parallel due to the need to update all members of the population, and it is hence ideally suited for the massively parallel hardware provided by GPUs. It is important to take into account that GPUs have properties of vector machines as well as features of parallel computers [32]. Certain groups of threads are scheduled for execution on the available multiprocessors simultaneously and operate in lockstep. Memory transactions for such thread groups are served collectively, and it is hence crucial to lay out data in memory in a way that ensures that logically neighboring threads access consecutive locations in GPU memory. It is possible to run the algorithm with a setup with only one thread per replica which is particularly simple and easily generalized to different systems. To achieve best performance, however, we used a checkerboard decomposition [33–35] of the lattice and hence several threads to update parts of each replica in parallel, for details see Ref. [31]. To achieve good memory *coalescence* of accesses for this setup, it is crucial to store spin configurations in a *spin-parallel* way, i.e., such that neighboring spins in memory correspond to neighboring lattice sites. Additionally, we store the two sub-lattices separately which improves the performance further. The total amount of data transferred over the GPU bus is kept at a minimum by using 8-bit integers for the spin variables and employing the stateless, counter-based random-number generator (RNG) Philox [36, 37]. To achieve decent performance on current devices it is vital to use a number of threads that is several times larger than the available number of actual compute units, as this allows the scheduler to pause thread groups waiting for data and activate other groups that have already completed memory accesses instead. The resulting benefit of *latency hiding* is fully taken advantage of in the present setup as typical population sizes $R$ are at least $10^4$ while the number of cores of present cards is of the order of $10^3$.

An implementation of the resampling process on GPU requires several steps. The normalization constants $Q_i$ of Eq. (1) are calculated using a parallel or "butterfly" reduction

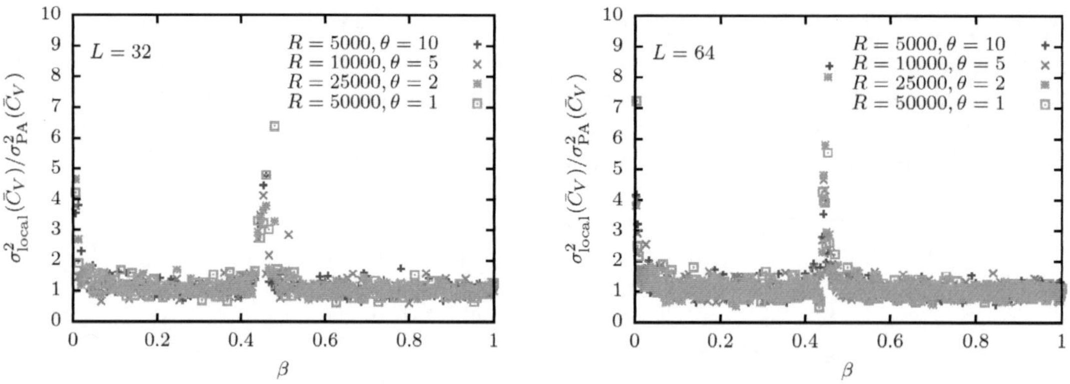


**Figure 1.** Ratio of variances of the specific heat of the 2D Ising model as estimated from simple Metropolis simulations of $R$ samples and $\theta$ sweeps and of PA simulations with the same parameters. The inverse temperature steps were $\Delta\beta = 0.0067$ for $L = 32$ and $\Delta\beta = 0.0033$ for $L = 64$. The protocols for both types of simulations are identical apart from the fact that for the Metropolis runs the resampling step was turned off.

method [31]. The resampling factors $\hat{\tau}_i(E_j)$ are then computed in a separate kernel according to the prescription in Eq. (3). The resampling itself is done in parallel on GPU too and leads to a memory requirement that is twice that of the actual population size. (Note that the population size $R_i$ weakly fluctuates between temperature steps.) Finally, measurements are again performed in a *replica-parallel* way for elementary quantities such as the energy and magnetization and their respective moments, using parallel reductions of the values for different population members to accumulate the population averages.

The resulting GPU code shows excellent performance as is illustrated by the peak performance data for the Ising case $q = 2$ collected for a Tesla K80 GPU and summarized in Table 1. The speedup of more than 200 times against a serial CPU code is almost independent of system size. For this specific case of only two states, an additional compression of spin states into the machine words is possible, leading to a method usually called *multi-spin coding* (MSC) [38] that yields additional performance increases against the standard *single-spin coded* (SSC) program. In the context of PA, this is most efficiently done by storing spins from the same lattice sites but different replicas in one word (sometimes called asynchronous multi-spin coding [35]). A performance bottle-neck in this setup is the generation of random numbers for the Metropolis updates. We use a linear-congruential generator seeded by the high-quality Philox RNG and achieve an about 10-fold speedup for MSC with 32 spins per word against the SSC implementation (see details in Ref. [31]). The corresponding spin-update times and speedups compared to the serial CPU code are summarized in Table 1.

## 4. Behavior at continuous transitions

The question of how well PA performs compared to other approaches, and in particular MCMC algorithms, can be answered quantitatively in terms of the dependence of the systematic errors (bias) and the statistical errors on the simulation parameters as well as on properties of the system such as the system size. We have discussed elsewhere the dependence of bias and variances on $R$, $\theta$ and $\Delta\beta$ for the example of the Ising model corresponding to the Hamiltonian (5) for $q = 2$ [25]. Here we focus on the computational efficiency in the vicinity of the critical point.

Figure 1 shows the ratio of estimated variances (squared error bars) for the specific heat of the 2D Ising model from two different algorithms: a Metropolis update and the PA method. Note

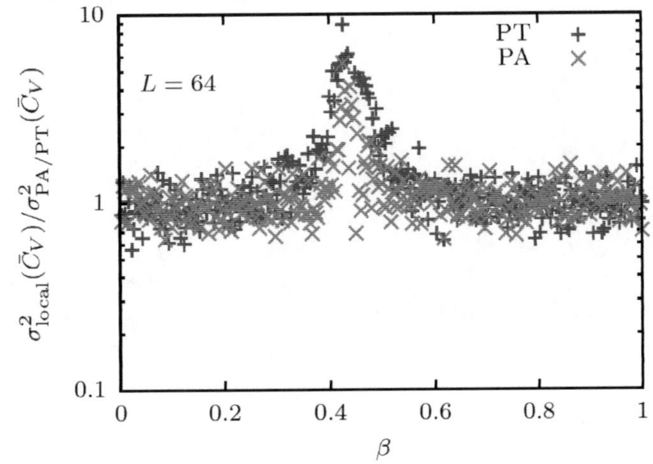


**Figure 2.** Ratio of variances of the specific heat for PA and parallel tempering (PT) simulations with the same number of MC steps as a function of inverse temperature $\beta$. In contrast to the data in Fig. 1, spin flips were performed according to a heatbath update rule here [14]. For PA we chose $R = 50\,000$ and $\theta = 1$ and used the same number of updates for the PT run.

that the Ising model at inverse temperature $\beta$ corresponds to the $q = 2$ Potts model at $2\beta$. Given that statistical errors in both methods decay with the inverse square root of the computational effort [14, 25] and that the run times (on CPU) of both setups were identical apart from a small overhead for the resampling steps, this ratio indicates an algorithmic speedup from using PA over a standard spin-flin only simulation, i.e., it shows how much longer the Metropolis simulation would need to run to produce error bars of the same size as those of the PA approach. To make the comparison as fair as possible, the type of Metropolis simulation used here is very similar to the setup in PA, i.e., a population of $R$ replicas is simulated in parallel starting from $\beta_0 = 0$ and decreasing the temperature in steps of $\Delta\beta$, performing $\theta$ Metropolis sweeps (and measurements) at each temperature. The Metropolis setup is hence identical to the PA algorithm, but only with the resampling step turned off. As Fig. 1 reveals, the resampling is irrelevant sufficiently far away from the critical point, resulting in a unit variance ratio. In this regime, the MCMC steps are sufficient to decorrelate the configurations at each temperature step and hence resampling cannot decrease statistical errors further. In contrast, in the vicinity of the phase transition at $\beta_c \approx 0.44068$ critical slowing down leads to an imperfect decorrelation of configurations through the MCMC moves, and in this case the resampling results in an additional relaxation of configurations. This yields reduced statistical errors for the PA runs there and consequently a speedup factor that is larger than one. The two panels in Fig. 1 for system sizes $L = 32$ and $L = 64$ show such a speedup, which appears to increase with system size.

Population annealing is similar in spirit to parallel tempering [12] and one might hence wonder how they compare to each other for simulating a system with a continuous transition. Figure 2 shows a comparison of the statistical speedup factors defined through the ratio of variances against the spin-flip only run for the PA simulation (corresponding to the data from the right panel of Fig. 1 for $R = 50\,000$) and a parallel tempering (PT) simulation using the same temperature sequence and the same number of spin updates. It is clear that both PT and PA show comparable speedup factors for this problem, an observation that is in line with experiences previously reported for spin-glass systems [13, 24]. At the same time, however, it is clear that PA is much better suited for highly parallel computational environments than PT.

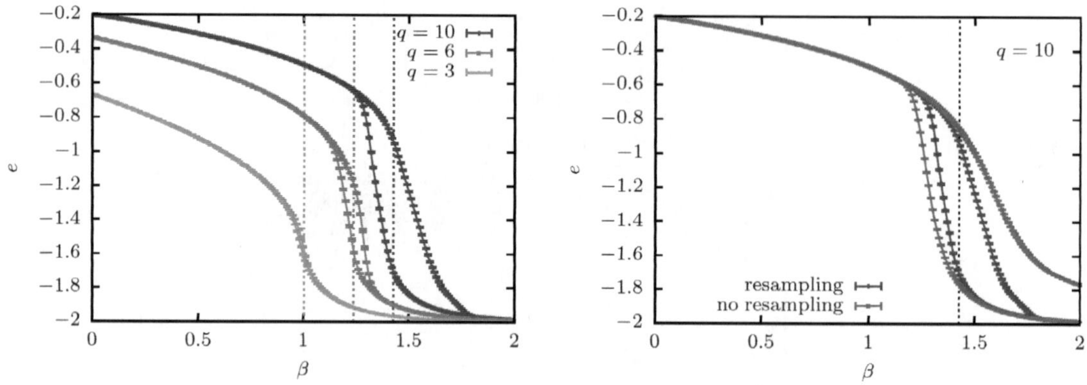


**Figure 3.** Hysteresis effect in PA runs for the $q$-states Potts model. Left: Energy as measured in cooling runs (right curve of each color) and heating runs (left curve of each color) for $q = 3$, $q = 6$ and $q = 10$ for $L = 32$ and $R = 10\,000$ with $\theta = 10$ and $\Delta\beta = 0.01$. The vertical lines indicate the asymptotic transition points at $\beta_t = \ln(1 + \sqrt{q})$. Right: The data points for $q = 10$ compared to the data for equivalent PA runs with the resampling step turned off.

## 5. Behavior at first-order transitions

At first-order points it is not the critical slowing down seen for continuous transitions that impedes the efficient simulation but rather the phenomenon of phase coexistence and the associated metastability [39]. As a consequence, the system remains in its original phase as the transition point is crossed and decays to the now stable opposite phase on a time scale that depends on the cooling or heating rate. This is the well-known effect of hysteresis. An established way of preventing this effect to occur is the simulation in generalized ensembles that artificially enhance the suppressed states connecting both phases, for example through the multicanonical method [16, 40]. It is interesting to see how the ensemble of configurations in PA behaves in this respect [41].

To clearly reveal any potential hysteresis effect, we added to the standard PA protocol that cools the population from its initial temperature point at $\beta_0$ down to $\beta_f$ a heating run, starting from $\beta_f$ and ending at $\beta_0$. The algorithm described in Sec. 2 works in the same way for such a temperature schedule, but now one needs to prepare the population in equilibrium at the inverse temperature $\beta_f$ instead of at $\beta_0 = 0$. This is achieved by preparing the population as a uniformly random sample of the $q$-fold degenerate ground state of the Hamiltonian (5), corresponding to an equilibrium sample at $T = 0$, i.e., at $\beta_f = \infty$. In practice, it is also an almost perfect approximation of the equilibrium distribution at the finite $\beta_f = 3$ used here.

The left panel of Fig. 3 shows the internal energies $e = \langle \mathcal{H} \rangle / N$ estimated from PA runs with a cooling and runs with a heating schedule, respectively, for various values of $q$. It is clearly visible that hysteresis occurs for the first-order cases $q = 6$ and $q = 10$ while it is absent for the second-order model $q = 3$. This hysteresis effect is found to be more pronounced for large (inverse) temperature steps $\Delta\beta$. Thus, PA does not remove the hysteresis observed in a purely local Metropolis simulation. Nevertheless, it is able to reduce this effect as is apparent from a comparison of the energy curves for the regular PA simulations to PA runs with the resampling step turned off, corresponding to pure Metropolis simulations. This is shown in the plot in the right panel of Fig. 3.

Such hysteresis is also observed in other quantities, for example in estimates of the free energy deduced from the expression of Eq. (4). This is illustrated in Fig. 4 for the same values of $q$ as shown in Fig. 3 and the additional case of the Ising model. Recall that in order to get absolute

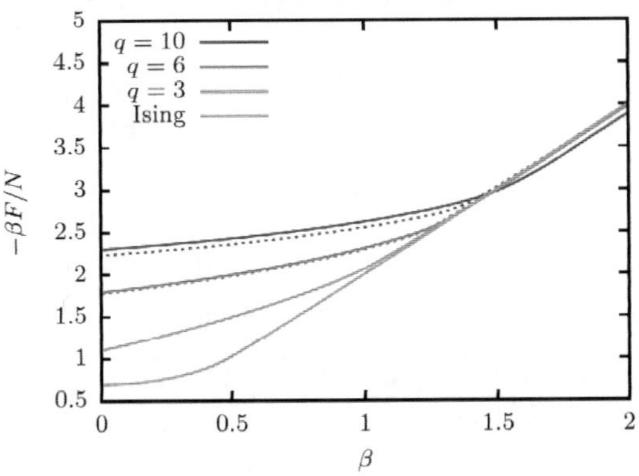


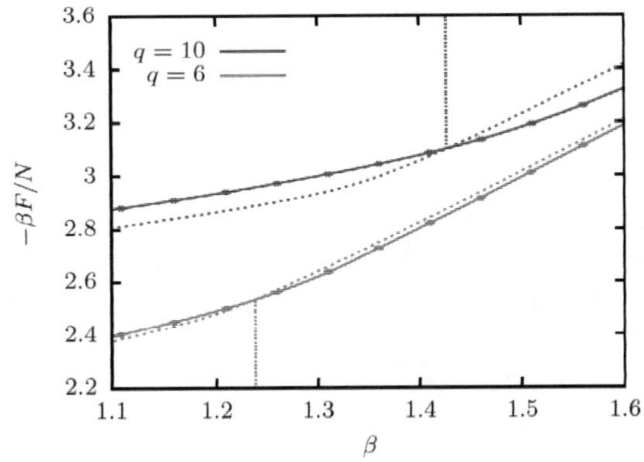


**Figure 4.** Left: Free energy of the Ising model and the $q = 3$, $q = 6$ and $q = 10$ Potts models as estimated from PA runs with cooling (solid lines) and with heating (dashed lines) schedules. The system size studied was $L = 32$ and the PA parameters were chosen to be $R = 10\,000$, $\theta = 10$ and $\Delta\beta = 0.01$. Right: Detail of the crossing of the metastable free energies for $q = 6$ and $q = 10$. Symbols with error bars are only shown for every fifth actual data point. The vertical dotted lines indicate the locations of the (exact) asymptotic transition points.

free energies, the expression (4) requires additional input in the form of the reference value $Z_{\beta_0}$. For the standard cooling schedule the relevant reference point is $-\beta F(\beta)/N = \ln q$ as $\beta \to 0$. For the heating runs, on the other hand, it is easy to see that $-\beta F(\beta)/N = (\ln q)/N - \beta e_0$ as $\beta \to \infty$, where $e_0 = E_0/N = -2$ is the ground-state energy per site [41]. The resulting free-energy estimates from cooling and from heating runs are perfectly consistent with each other for the second-order cases $q = 2$ (Ising) and $q = 3$, cf. the left panel of Fig. 4. For the first-order models $q = 6$ and $q = 10$, on the other hand, the low-temperature and high-temperature estimates differ, as expected [40], and both curves intersect close to the transition point, thus providing a method for determining from the simulations the location of the phase transition [40, 41].

## 6. Summary

We have studied population annealing as a promising new technique for the simulation of systems undergoing phase transitions. Being a hybrid of sequential and Markov chain Monte Carlo methods, it has the advantage of a theoretically perfect strong scaling behavior. While on increasing the number of parallel workers pure Markov chain methods eventually spend most time in the equilibration phase, bias as well as statistical errors in population annealing are systematically reduced with increasing numbers of parallel threads. Taking advantage of these excellent scaling properties we presented an efficient GPU implementation of population annealing for the Potts model. For the Ising case $q = 2$ we find a more than 200-fold speedup against a serial CPU code even for small system sizes. This can be further increased by using multi-spin coding techniques which yield an additional factor of 10 for the Ising model. Population annealing is found to offer no algorithmic advantage over the underlying Markov chain technique in terms of a possible reduction of statistical errors from the same computational effort for the Ising model off criticality. This is where the regular spin flips are sufficient to decorrelate configurations. In the vicinity of the critical point, on the other hand, an algorithmic speedup is achieved that potentially grows with system size. For systems with first-order phase transitions, population annealing is able to reduce, but not to eliminate the effect of hysteresis caused by metastability close to the transition. The free-energy estimate naturally deduced from population annealing runs can be used to estimate the location of the transition point. A combination of population annealing with generalized-ensemble methods that might be able to more completely remove the effects of metastability would be an interesting topic for future research.

**Acknowledgments**
The work of L.B., M.B. and L.S. is supported by grant No. 14-21-00158 of the Russian Science Foundation. M.B. was also supported by the Scientific Grant Agency of the Ministry of Education of the Slovak Republic (Grant No. 1/0331/15). The authors acknowledge support from the European Commission through the IRSES network DIONICOS under Contract No. PIRSES-GA-2013-612707.

**References**

[1] Iba Y 2001 *Trans. Jpn. Soc. Artif. Intell.* **16** 279–286
[2] Hukushima K and Iba Y 2003 *AIP Conf. Proc.* **690** 200–206
[3] Ceperley D M and Kalos M H 1986 Quantum many-body problems *Monte Carlo Methods in Statistical Physics* ed Binder K (Berlin: Springer) pp 145–194
[4] Kitagawa G 1996 *J. Comp. Graph. Stat.* **5** 1–25
[5] Grassberger P 2002 *Comput. Phys. Commun.* **147** 64–70
[6] Doucet A, de Freitas N and Gordon N (eds) 2001 *Sequential Monte Carlo Methods in Practice* (New York: Springer)
[7] Asanovic K, Bodik R, Catanzaro B C, Gebis J J, Husbands P, Keutzer K, Patterson D A, Plishker W L, Shalf J, Williams S W and Yelick K A 2006 The landscape of parallel computing research: A view from Berkeley *Tech. Rep. University of California, Berkeley, EECS Department, UCB/EECS-2006-183*
[8] Owens J D, Houston M, Luebke D, Green S, Stone J E and Phillips J C 2008 *Proceedings of the IEEE* **96** 879–899
[9] Zierenberg J, Marenz M and Janke W 2013 *Comput. Phys. Commun.* **184** 1155–1160
[10] Vogel T, Li Y W, Wüst T and Landau D P 2013 *Phys. Rev. Lett.* **110** 210603
[11] Gross J, Zierenberg J, Weigel M and Janke W 2017 Parallel multicanonical simulations on GPUs *Preprint* in preparation
[12] Machta J and Ellis R S 2011 *J. Stat. Phys.* **144** 541–553
[13] Wang W, Machta J and Katzgraber H G 2015 *Phys. Rev.* E **92** 063307
[14] Landau D P and Binder K 2015 *A Guide to Monte Carlo Simulations in Statistical Physics* 4th ed (Cambridge: Cambridge University Press)
[15] Swendsen R H and Wang J S 1987 *Phys. Rev. Lett.* **58** 86–88
[16] Berg B A and Neuhaus T 1992 *Phys. Rev. Lett.* **68** 9–12
[17] Wang F and Landau D P 2001 *Phys. Rev. Lett.* **86** 2050–2053
[18] Hukushima K and Nemoto K 1996 *J. Phys. Soc. Jpn.* **65** 1604–1608
[19] Wang W, Machta J and Katzgraber H G 2014 *Phys. Rev.* B **90** 184412
[20] Wang W, Machta J and Katzgraber H G 2015 *Phys. Rev.* B **92** 094410
[21] Machta J 2010 *Phys. Rev.* E **82** 026704
[22] Ferrenberg A M and Swendsen R H 1988 *Phys. Rev. Lett.* **61** 2635–2638
[23] Garel T and Orland H 1990 *J. Phys.* A **23** L621–L626
[24] Wang W, Machta J and Katzgraber H G 2015 *Phys. Rev.* E **92** 013303

[25] Weigel M, Barash L Y, Borovský M, Shchur L N and Janke W 2017 Population annealing for the Ising model *Preprint* in preparation

[26] Wu F Y 1982 *Rev. Mod. Phys.* **54** 235–268

[27] Baxter R J 1973 *J. Phys.* C **6** L445–L448

[28] Beffara V and Duminil-Copin H 2012 *Probab. Theory Rel.* **153** 511–542

[29] Duminil-Copin H, Sidoravicius V and Tassion V 2017 *Commun. Math. Phys.* **349** 47–107

[30] Duminil-Copin H, Gagnebin M, Harel M, Manolescu I and Tassion V 2016 Discontinuity of the phase transition for the planar random-cluster and Potts models with $q > 4$ *Preprint* arXiv:1611.09877

[31] Barash L Y, Weigel M, Borovský M, Janke W and Shchur L N 2017 GPU accelerated population annealing algorithm *Preprint* arXiv:1703.03676 *Comput. Phys. Commun.* in print (http://dx.doi.org/10.1016/j.cpc.2017.06.020)

[32] Kirk D B and Hwu W W 2010 *Programming Massively Parallel Processors* (Amsterdam: Elsevier)

[33] Weigel M 2012 *J. Comp. Phys.* **231** 3064–3082

[34] Weigel M 2011 *Comput. Phys. Commun.* **182** 1833–1836

[35] Lulli M, Bernaschi M and Parisi G 2015 *Comput. Phys. Commun.* **196** 290–303

[36] Salmon J K, Moraes M A, Dror R O and Shaw D E 2011 Parallel random numbers: As easy as 1, 2, 3 *Proceedings of 2011 International Conference for High Performance Computing, Networking, Storage and Analysis* SC '11 (New York: ACM)

[37] Manssen M, Weigel M and Hartmann A K 2012 *Eur. Phys. J. Special Topics* **210** 53–71

[38] Zorn R, Herrmann H J and Rebbi C 1981 *Comput. Phys. Commun.* **23** 337–342

[39] Binder K 1987 *Rep. Prog. Phys.* **50** 783–859

[40] Janke W 2003 First-order phase transitions *Computer Simulations of Surfaces and Interfaces* (*NATO Science Series, II. Mathematics, Physics and Chemistry* vol 114) ed Dünweg B, Landau D P and Milchev A I (Dordrecht: Kluwer Academic Publishers) pp 111–135

[41] Barash L Y, Weigel M, Shchur L N and Janke W 2017 *Eur. Phys. J. Special Topics* **226** 595–604

IOP Conf. Series: Journal of Physics: Conf. Series **921** (2017) 012018 doi:10.1088/1742-6596/921/1/012018

# Transition barrier at a first-order phase transition in the canonical and microcanonical ensemble

**Wolfhard Janke[1], Philipp Schierz[1] and Johannes Zierenberg[1,2,3]**

[1] Institut für Theoretische Physik, Universität Leipzig, Postfach 100 920, 04009 Leipzig, Germany

[2] Max Planck Institute for Dynamics and Self-Organization, Am Fassberg 17, 37077 Göttingen, Germany

[3] Bernstein Center for Computational Neuroscience, Am Fassberg 17, 37077 Göttingen, Germany

E-mail: wolfhard.janke@itp.uni-leipzig.de

**Abstract.** We compare the transition barrier that accompanies a first-order phase transition in the canonical and microcanonical ensemble. This is directly encoded in the probability distributions of standard Metropolis Monte Carlo simulations and a proper microcanonical sampling technique. For the example of droplet formation, we find that in both ensembles the transition barrier scales as expected but that the barrier is much smaller in the microcanonical ensemble. In addition its growth with system size is weaker which will enhance this difference for larger systems. We provide an intuitive physical explanation for this observation.

## 1. Introduction

Many relevant transitions in nature fall into the class of first-order phase transitions, where directly at the transition point two or more phases coexist, separated by highly suppressed transition states. Prime examples of first-order transitions include the gas-liquid transition or the change of magnetic order under variation of an external field.

The formation of a droplet in a supersaturated gas is also a first-order phase transition, where instead of coexistence of two pure phases (e.g., gas and liquid) one observes coexistence between the homogeneous gas phase and a phase with a single droplet in equilibrium with the surrounding vapor. While the basic properties of this nucleation process have been thoroughly discussed [1, 2, 3, 4, 5], many puzzling details remain open. In fact, nucleation rates predicted by simulations and measured in experiments still do not match. Classical nucleation theory connects the rate $R$ of droplet formation with the free-energy barrier $\Delta F$: $R = \kappa e^{-\beta\Delta F}$. The kinetic prefactor $\kappa$ includes the kinetic details of the nucleation process and the free-energy barrier may be related to the suppression in the probability distribution of a suitable reaction coordinate. Both may be, in principle, computed from Monte Carlo simulations [6, 7]. This notion is commonly adapted from the canonical ensemble and we hence refer to the free-energy barrier $\beta\Delta F$ as a general transition barrier $B$ in the following.

In case of the condensation-evaporation transition both the droplet size and the potential energy are suitable reaction coordinates. We focus here on the potential-energy probability distribution $P(E_p)$ and immediately notice that this strongly depends on the thermodynamic ensemble. If the system is in a heat bath (canonical ensemble) there is an exchange of energy with the surrounding; if the system is isolated (microcanonical ensemble) then there is no energy flux. Of course, in the latter case the system may transfer kinetic to potential energy and vice versa, which allows for a well-defined potential-energy probability distribution.

In the following, we will discuss the effect of the ensemble on the transition barrier of a general first-order phase transition for which the energy is a suitable reaction coordinate. In Sec. 2 we describe a proper microcanonical sampling technique that allows us to obtain the potential-energy probability distribution in both ensembles. Combined with multi-histogram reweighting techniques, we get direct access to the suppression of transition states – the transition barrier. The finite-size scaling is discussed in Sec. 3 followed by an intuitive explanation of the somewhat surprising results. We finish with our conclusions in Sec. 4.

## 2. Method

We employ Monte Carlo simulations in the "real" microcanonical ensemble [8, 9, 10, 11, 12], referring to the conservation of total energy $E$ which is the standard textbook definition of the microcanonical ensemble. We emphasize this since most previous applications of the microcanonical ensemble in Monte Carlo simulations have focused on the conservation of potential energy $E_p$. The reason for this will become clear when we briefly discuss the method in the following. In addition to the conservation of the total energy $E$, we further fix the particle number $N$ and the volume $V$. This defines the NVE ensemble with the partition function $\Omega(E) = \int\int \mathcal{D}x\ \mathcal{D}p\ \delta(E - [E_p(x) + E_k(p)])$, where $\mathcal{D}x$ denotes the integration over state space and $\mathcal{D}p$ over momentum space. Integrating out the momentum degrees of freedom, which enter the kinetic energy $E_k = \sum_i p_i^2/2m$, this can be reduced to the (restricted) potential-energy space. For $N$ particles in three dimensions one obtains [8]

$$\Omega(E) = \frac{(2\pi m)^{\frac{3N}{2}}}{\Gamma(\frac{3N}{2})} \int_{-\infty}^{\infty} dE_p \hat{\Omega}(E_p)(E - E_p)^{\frac{3N-2}{2}} \Theta(E - E_p), \tag{1}$$

where $\Gamma(n)$ is the Gamma function, $\hat{\Omega}(E_p)$ is the (conformational) density of states and $\Theta(E - E_p)$ is the Heaviside step function reflecting the constraint $E_p \leq E$.

We can thus sample the microcanonical phase space by generating a Markov chain according to the weight

$$W_{\mathrm{NVE}}(E_p) = (E - E_p)^{\frac{3N-2}{2}}\, \Theta(E - E_p)\,, \tag{2}$$

where of course we have to start from a potential energy $E_p < E$ since $E_p > E$ has zero probability in this ensemble. The usual Metropolis acceptance probability for a proposed move from (micro)state $A$ to $B$ is then naturally adapted to the NVE ensemble:

$$P_{\mathrm{acc}}\,(A \to B) = \min\left\{1, W_{\mathrm{NVE}}(E_p^B)/W_{\mathrm{NVE}}(E_p^A)\right\}. \tag{3}$$

Here one clearly sees the difference to fixing the potential energy in a conformational microcanonical ensemble. While this might be a natural ensemble for spin systems, where a kinetic contribution is not properly defined (but may be exploited for numerical purposes [9]), it is an incomplete ensemble for general systems in soft condensed matter. Of course, a standard Monte Carlo approach in the canonical ensemble makes use of this reduction. This is valid for all canonical expectation values that are independent of the kinetic contributions. However, if one considers quantities that are derived from energy probability distributions, e.g., the free-energy barrier, the contributions of the kinetic energy turn out to play a role [7].

The importance sampling defined in (3) can be combined with a replica-exchange scheme, where parallel simulations at different total energies exchange their configurations with the probability

$$P_{\mathrm{exc}}\,(A \leftrightarrow B) = \min\left\{1, \frac{W_{\mathrm{NVE^B}}(E_p^A) W_{\mathrm{NVE^A}}(E_p^B)}{W_{\mathrm{NVE^B}}(E_p^B) W_{\mathrm{NVE^A}}(E_p^A)}\right\}. \tag{4}$$

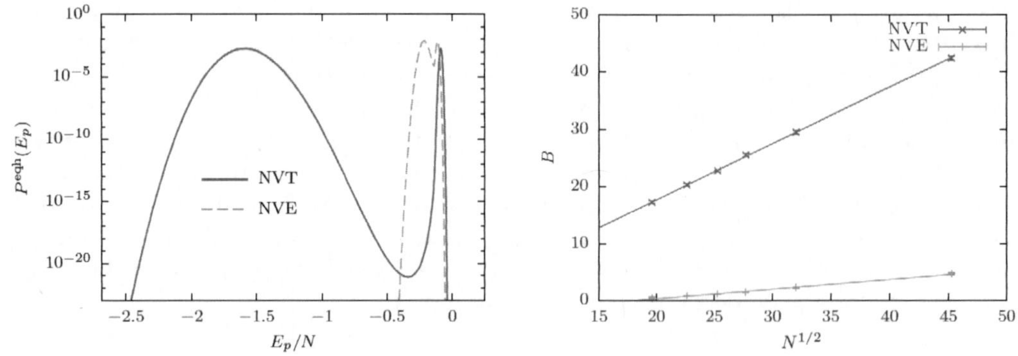


**Figure 1.** (left) Probability distributions of the potential energy in the canonical (NVT) and microcanonical (NVE) ensemble for 2048 Lennard-Jones particles at fixed density $\rho = 10^{-2}$. The distributions are obtained at the equal-height temperature $T_{\rm eqh} = 0.6175$ and equal-height total energy $E_{\rm eqh}/N = 0.6849$, respectively. (right) The free-energy barrier of the condensation-evaporation transition in a Lennard-Jones system for the (conformational) NVT and NVE ensembles. This is directly related to the sampling barrier in the respective importance sampling schemes.

Afterwards, NVE WHAM can be applied to estimate the density of states $\hat{\Omega}(E_p)$ [10, 13, 14, 15]. This is an ensemble-independent property of the simulated system and allows one to estimate observables in other ensembles, e.g., the canonical NVT ensemble [10, 16].

## 3. Results

As an illustrative example we show results for the 12–6 Lennard-Jones particle system where the particles $i$ and $j$ interact via the potential $V_{\rm LJ}(r_{ij}) = 4\epsilon\left[(\sigma/r_{ij})^{12} - (\sigma/r_{ij})^{6}\right]$, cutoff and shifted at $r_{ij} = 2.5\sigma$, with $\epsilon = 1$ setting the energy (or temperature) and $\sigma = 2^{-1/6}$ setting the length scale of the system. We consider $N$ particles in a cubic box of volume $V$ with periodic boundary conditions and fix the density to $\rho = N/V = 0.01$. To update the particle positions, we apply short-range displacement and long-range jump proposals, with symmetric selection probabilities each. The combination of replica-exchange NVE sampling and NVE WHAM allows us to directly estimate the potential-energy probability distribution in both the canonical ensemble, $P_{\rm NVT}(E_p) \propto \hat{\Omega}(E_p)\exp(-E_p/k_BT)$, and microcanonical ensemble, $P_{\rm NVE}(E_p) \propto \hat{\Omega}(E_p)W_{\rm NVE}(E_p)$. Both show, as expected, a double peak at the condensation-evaporation transition, see Fig. 1 (left). In the canonical ensemble, we have to vary the temperature to obtain a wide distribution with pronounced peaks of equal height at $T_{\rm eqh} = 0.6175$. In the microcanonical ensemble, we vary the total energy to obtain a comparably narrow double-peak distribution at $E_{\rm eqh}/N = 0.6849$.

In addition to being much narrower, we also notice that the suppression of transition states in the microcanonical probability distribution is much lower than in the canonical distribution. This corresponds to a much lower barrier $B_{\rm NVE} \ll B_{\rm NVT}$, each defined by

$$B = \ln\left[P^{\rm eqh}\left(E_p^{\pm}\right)/P^{\rm eqh}\left(E_p^{0}\right)\right], \tag{5}$$

where $P^{\rm eqh}(E_p)$ is the (equal-height) potential-energy distribution in the respective ensemble, $E_p^{\pm}$ refers to the location of the two maxima and $E_p^{0}$ to the location of the minimum in between. Figure 1 (right) shows the canonical and microcanonical barriers for system sizes $N \in \{384, 512, 640, 768, 1024, 2048\}$. The microcanonical barrier is always below the canonical

IOP Conf. Series: Journal of Physics: Conf. Series **921** (2017) 012018 doi:10.1088/1742-6596/921/1/012018

barrier. In fact, the suppression is several orders of magnitude smaller as best illustrated for the $N = 2048$ system where $B_{\mathrm{NVT}} = 42.4$ and $B_{\mathrm{NVE}} = 4.7$ – recall that this measures the logarithm of the suppression. In this case, importance sampling in the NVE ensemble is of the order of $10^{16}$ times more efficient than in the canonical counterpart for this transition. This observation can be derived from the ensemble weights [11] and thus is generally true for any first-order phase transition with suppressed transition states in the energy probability distribution. It can be further generalized to any properly defined probability distribution and tailored ensemble weight.

From the canonical ensemble we know that the barrier is related to the surface of the interface that separates the coexisting phases, here the liquid droplet from the surrounding gas. We expect that the barrier is thus proportional to the surface of the droplet, $B \propto \partial V_D$. The surface depends on the droplet volume, $\partial V_D \propto V_D^{2/3}$. Due to the interplay of energy minimization inside the droplet and entropy maximization in the surrounding gas, the droplet volume is expected to grow with system size as $V_D \propto V^{3/4}$ [17, 18, 19, 20], or at fixed density as $V_D \propto N^{3/4}$ [21]. Introducing an effective interfacial free energy $\tau$ and considering additional logarithmic corrections [22, 23, 24, 25, 26] we arrive at the finite-size scaling ansatz in the canonical ensemble [7]

$$B = \tau N^{1/2} - \alpha \ln N + c, \tag{6}$$

where $\alpha$ and $c$ are constants.

In the microcanonical ensemble, the setup directly at the transition energy is very similar to the canonical ensemble: A fixed volume and particle number, while the system switches between a homogeneous gas and a droplet in coexistence with the surrounding vapor. In fact, the conservation of total energy introduces a potential-energy reservoir from which the system can take potential energy to form a droplet or where it can store it to form a gas. In a limited scope, this is locally comparable to the canonical ensemble. We hence assume the same dependence of the barrier on the droplet surface, which trivially scales with the droplet volume. The relation between droplet size and system size is as well assumed to be consistent with the canonical ensemble, assuming the energy and entropy arguments to be transferable. We thus assume the same scaling ansatz (6).

Figure 1 (right) shows the canonical and microcanonical barriers which satisfy to leading order the scaling ansatz (6). For simplicity, we only consider a few system sizes and restrict fits to the effective behavior $B = \tau N^{1/2} + c$, which yields $\tau_{\mathrm{NVT}} = 0.98(2)$ and $\tau_{\mathrm{NVE}} = 0.166(8)$, with goodness-of-fit parameters $Q \approx 0.79$ and $Q \approx 0.18$, respectively. Please note that we here focus on the conformational canonical ensemble because of the direct relation to the Monte Carlo sampling methods. Working with the full canonical ensemble, however, only shifts the line but leaves the $\tau$ estimate invariant [7].

Let us finally give a physical picture that heuristically explains the difference of the sampling behavior in the NVT and NVE ensembles for the droplet condensation-evaporation transition. In the NVT ensemble, the system is in a heat bath, which ensures that there is always energy available to be added to the system. The result is a constant transition probability to higher potential-energy configurations with the same energy step $\Delta E_p$ independent of the specific value of the potential energy. At the canonical transition temperature, this allows in principle to transform the coexisting (possibly large) droplet at low potential energy into a gas with high potential energy. The situation is different in the NVE ensemble, where a distinct amount of energy is available and distributed in potential and kinetic energy. The potential energy may only be raised by transferring kinetic energy with decreasing probability until none is left. This puts an upper bound on the reachable potential energies. The result is a decreasing transition probability to higher potential-energy configurations which therefore depends on the current potential energy and goes to zero when approaching the upper bound. At the point where the energy versus entropy competition evens out in the NVT ensemble, the NVE ensemble will still be stuck in the droplet phase since higher energies show a higher suppression. Consequently, phase coexistence occurs in the NVE ensemble at a smaller droplet size, which explains the smaller barrier.

## 4. Conclusions

We demonstrated on the example of the condensation-evaporation transition that the transition barrier of a first-order phase transition is always smaller in the microcanonical ensemble compared to the canonical ensemble. In fact, a finite-size scaling analysis reveals that this effect drastically increases in the limit of increasing system sizes. A physical reason for this is that in a heat bath (canonical ensemble) the system can in principle increase its potential energy infinitely, because there is an infinite reservoir of potential energy – the heat bath. In the microcanonical ensemble, the finite reservoir of potential energy due to the constraint of a fixed total energy allows only very narrow distributions with a reduced barrier [11]. These general statements about the ensembles are not restricted to computational considerations. Physical microcanonical signatures might be approximately observed in astrophysics or extremely isolated systems on earth where the barrier difference should show up as well.

## Acknowledgments

This work has been partially supported by the DFG (Grant No. JA 483/31-1), the Leipzig Graduate School "BuildMoNa", and the Deutsch-Französische Hochschule DFH-UFA (Grant No. CDFA-02-07). The authors gratefully acknowledge the computing time provided by the John von Neumann Institute for Computing (NIC) on the supercomputer JURECA at Jülich Supercomputing Centre (JSC) under Grant No. HLZ24. JZ received financial support from the German Ministry of Education and Research (BMBF) via the Bernstein Center for Computational Neuroscience (BCCN) Göttingen under Grant No. 01GQ1005B.

## References

[1] Volmer M and Weber A 1925 *Z. Phys. Chem.* **119** 277
[2] Becker R and Döring W 1935 *Ann. Phys.* **416** 719
[3] Feder J, Russell K C, Lothe J and Pound G M 1966 *Adv. Phys.* **15** 111
[4] Oxtoby D W 1992 *J. Phys: Condens. Matter* **4** 7627
[5] Kashchiev D 2000 *Nucleation: Basic Theory with Applications* (Oxford: Butterworth-Heinemann)
[6] Auer F and Frenkel D 2001 *Nature* **409** 1020
[7] Zierenberg J, Schierz P and Janke W 2017 *Nat. Commun.* **8** 14546
[8] Calvo F, Neirotti J P, Freeman D L and Doll J D 2000 *J. Chem. Phys.* **112** 10350
[9] Martin-Mayor V 2007 *Phys. Rev. Lett.* **98** 137207
[10] Schierz P, Zierenberg J and Janke W 2015 *J. Chem. Phys.* **143** 134114
[11] Schierz P, Zierenberg J and Janke W 2016 *Phys. Rev.* E **94** 021301(R)
[12] Davis S and Gutiérrez G 2017 *arXiv*:1703.05992
[13] Ferrenberg A M and Swendsen R H 1989 *Phys. Rev. Lett.* **63** 1195
[14] Kumar S, Rosenberg J M, Bouzida D, Swendsen R H and Kollman P A 1992 *J. Comput. Chem.* **13** 1011
[15] Kim J, Straub J E and Keyes T 2012 *J. Phys. Chem.* B **116** 8646
[16] Escobedo F A 2006 *Phys. Rev.* E **73** 056701
[17] Biskup M, Chayes L and Kotecký R 2002 *Europhys. Lett.* **60** 21
[18] Biskup M, Chayes L and Kotecký R 2003 *Commun. Math. Phys.* **242** 137
[19] Neuhaus T and Hager J S 2003 *J. Stat. Phys.* **113** 47
[20] Binder K 2003 *Physica* A **319** 99
[21] Zierenberg J and Janke W 2015 *Phys. Rev.* E **92** 012134
[22] Langer J S 1967 *Ann. Phys. N.Y.* **41** 108
[23] Ryu S and Cai W 2010 *Phys. Rev.* E **81** 030601(R)
[24] Nußbaumer A, Bittner E and Janke W 2010 *Prog. Theor. Phys. Suppl.* **184** 400
[25] Prestipino S, Laio A and Tosatti E 2012 *Phys. Rev. Lett.* **108** 225701
[26] Prestipino S, Laio A and Tosatti E 2014 *J. Chem. Phys.* **140** 094501

# Density of states for systems with multiple order parameters: a constrained Wang-Landau method

**Chor-Hoi Chan, Gregory Brown and Per Arne Rikvold**
Department of Physics, Florida State University, Tallahassee, FL 32306-4350, USA
E-mail: seahoi2001@gmail.com, gbrown@fsu.edu, prikvold@fsu.edu

**Abstract.** A macroscopically constrained Wang-Landau Monte Carlo method was recently proposed to calculate the joint density of states (DOS) for systems with multiple order parameters. Here we demonstrate results for a nearest-neighbor Ising antiferromagnet with ferromagnetic long-range interactions (a model spin-crossover material). Its two relevant order parameters are the magnetization $M$ and the staggered magnetization $M_s$. The joint DOS, $g(E, M, M_s)$ where $E$ is the total system energy, is calculated for zero external field and long-range interaction strength, and then obtained for arbitrary values of these two field-like model parameters by a simple transformation of $E$. Illustrations are shown for several parameter sets.

## 1. Motivation, method and model

To obtain phase diagrams for many-particle systems with multiple order parameters and field-like system parameters by standard importance sampling Monte Carlo (MC) simulations, one typically needs to perform simulations for a range of temperatures and field-like parameters. This is a quite tedious and computationally intensive task. Much effort can often be saved by calculating the density of states (DOS) in energy space, $g(E)$, using the Wang Landau (WL) method [1, 2] or one of its many later refinements. However, if a *joint* DOS involving the energy and several order parameters, $g(E, O_1, O_2, ...)$ is needed, the random walks that lie at the heart of the WL method become multi-dimensional and consequently quite slow [3].

Motivated by a computational problem of this nature (see specifics below), we recently developed an algorithm based on the WL method, in which the multidimensional random walk in energy and order parameters is replaced by many independent, one-dimensional walks in energy space, each constrained to fixed values of the order parameters [3]. For this reason, we call it the macroscopically constrained Wang-Landau (MCWL) method. (A similar, but less general, method was recently proposed by Ren *et al.* [4].) Our implementation of the algorithm is designed to execute in trivially parallel fashion on a large set of independent processors. Through further, symmetry based simplifications [3], the method provides an accurate estimate of a joint DOS with three variables in a relatively short time. All the simulations necessary to obtain the main numerical results presented here and in Refs. [3] and [5] were finished within one week, running independently on about two hundred separate computing cores. Here, we use the method to obtain the densities of states of a model spin-crossover material with two macroscopic order parameters, needed to produce phase diagrams under arbitrary external conditions. However, it has the potential to be applied to other complicated systems and phase spaces of higher dimension. Details of the algorithm and its implementation are given in Ref. [3].

The problem that inspired our development of the MCWL method is a model spin-crossover material. These materials are molecular crystals composed of magnetic ions surrounded by organic ligands. The molecules can be in two different magnetic states, a highly degenerate, high-spin (HS) and a less degenerate, low-spin (LS) state. The HS state has higher energy and larger molecular volume than the LS state, and the volume difference leads to elastic distortions of crystals with a mixture of HS and LS molecules. A simple model of phase transitions in some spin-crossover materials can be constructed as an antiferromagnetic Ising-like model with elastic distortions [6]. The elastic interactions can be further approximated as a ferromagnetic mean-field (equivalent neighbor or Husimi-Temperley) interaction, leading to the following Hamiltonian on an $L \times L$ square lattice with periodic boundary conditions [3, 7, 8],

$$\mathcal{H} = J \sum_{\langle i,j \rangle} s_i s_j - HM - \frac{A}{2L^2} M^2 , \quad (1)$$

with $J > 0$. Each spin, $s_i$ is either up (+1, HS) or down (−1, LS). $J \sum_{\langle i,j \rangle} s_i s_j$ represents the short-range interaction between neighboring spins, which favors adjacent spins aligning alternately up and down in an antiferromagnetic (AFM) configuration. $M = \sum_i s_i$ is the magnetization and $H$ is the strength of the applied field (which is actually an effective field in the spin-crossover material [8, 9]). The term $-HM$ favors all spins aligned with $H$ in the same direction. $L^2$ is the total number of sites, and $A \geq 0$ is the long-range interaction strength. Two spins located far apart on the lattice will interact through the term $M^2$, such that $-\frac{A}{2L^2}M^2$ has its most negative value if all the spins are aligned in the same direction. Therefore, this amounts to a ferromagnetic (FM) long-range interaction. We will refer to this model as the two-dimensional Ising Antiferromagnetic Short-range and Ferromagnetic Long-range (2D Ising-ASFL) model. Its two field-like system parameters are $H$ and $A$.

Throughout the paper, the total energy ($E$), magnetic field ($H$), and long-range interaction ($A$), will be expressed in units of $J$. The lattice is divided into two interpenetrating sublattices, A and B in a checkerboard fashion, with sublattice magnetizations $M_\mathrm{A}$ and $M_\mathrm{B}$, respectively. The two order parameters of the model are the magnetization ($M = M_\mathrm{A} + M_\mathrm{B}$) and the staggered magnetization ($M_\mathrm{s} = M_\mathrm{A} - M_\mathrm{B}$).

The conditional DOS, $g(E|M, M_\mathrm{s})$, is determined through independent, one-dimensional WL simulations for fixed $M$ and $M_\mathrm{s}$. The order parameters are kept constant by performing microscopic spin-exchange (Kawaski) updates separately on each sublattice. The joint DOS is then obtained by an analogue of Bayes' theorem as

$$g(E, M, M_\mathrm{s}) = \frac{g(E|M, M_\mathrm{s})}{\sum_E g(E|M, M_\mathrm{s})} g(M, M_\mathrm{s}) , \quad (2)$$

where $g(M, M_\mathrm{s})$ is determined by exact combinatorial calculation. It is only necessary to calculate $g(E|M, M_\mathrm{s})$ for one single set of $H$ and $A$. Here we use $H = A = 0$, which corresponds to a simple square-lattice Ising antiferromagnet [10]. The joint DOS for any arbitrary value of $(H, A)$ can be obtained by the simple transformation of $E$,

$$g(E, M, M_\mathrm{s}) \to g(E - HM - \frac{AM^2}{2L^2}, M, M_\mathrm{s}) . \quad (3)$$

This result is based on the fact that all microstates are equally shifted in energy when a field-like parameter, such as $H$ or $A$, couples to a function of a global property, such as $M$ (or $M_\mathrm{s}$), as shown in Eq. (1).

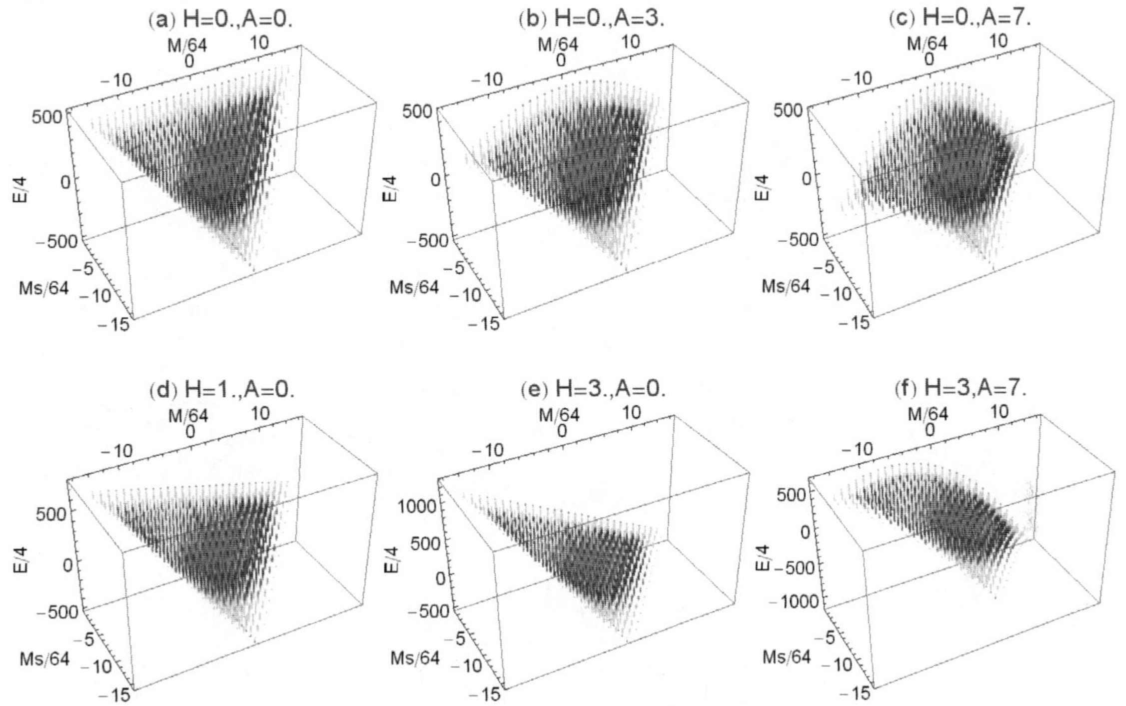


**Figure 1.** Natural logarithm of the joint DOS, $\ln g(E, M, M_s)$, of the 2D Ising-ASFL model for system size $L = 32$ for different field strengths ($H$) and long-range interaction strengths ($A$). The three axes are $M/64$, $M_s/64$, and $E/4$. The colors of the data points show the relative magnitude of the natural logarithm of the DOS, ranging from red (lowest) to magenta (highest). Only results for $M_s \leq 0$ are shown as there is reflection symmetry about the $M_s = 0$ plane. $M$ and $M_s$ are both sampled with an increment of 64. All possible energy levels are included, giving an increment of 4 for $E$ at fixed $M$ and $M_s$.

## 2. Results

Figure 1 shows the joint DOS, $g(E, M, M_s)$, for different magnetic fields $H$ and long-range interaction strengths $A$. Simulations were only performed for the case $H = 0$ and $A = 0$, shown in (a). Only 1/8 of the data points in (a) need to be calculated directly by simulation. The remaining data points are obtained through symmetry transformations in the $(M, M_s)$ plane as described in [3]. Data points in (b) to (f) are all obtained through transformation of the energy axis in (a), using Eq. (3). When the field $H$ is applied to the system as in (d) and (e), the term $-HM$ in the Hamiltonian pulls down the states with positive magnetization ($M > 0$) linearly, while it pulls up the states with negative magnetization ($M < 0$). On the other hand, with the long-range interaction ($A$) present, due to the term $-AM^2/2L^2$, states with non-zero magnetization ($M$) are pulled down in a quadratic manner. If both $H$ and $A$ are applied, some states are shifted up, and some are shifted down as shown in (f). Figure 1 shows results for the largest system size we studied, $L = 32$. While Fig. 1 shows the DOS versus three variables $(E, M, M_s)$, Figs. 2 and 3 show the DOS after summing the results along one axis, i.e., when the DOS is plotted against two variables only, $E$ and $M$, and $E$ and $M_s$, respectively.

Figure 2 shows $g(E, M)$. A similar effect as discussed for Fig. 1 is seen. I.e., when $H$ is applied, the energies for states with positive $M$ are pulled down, and the energies for states with

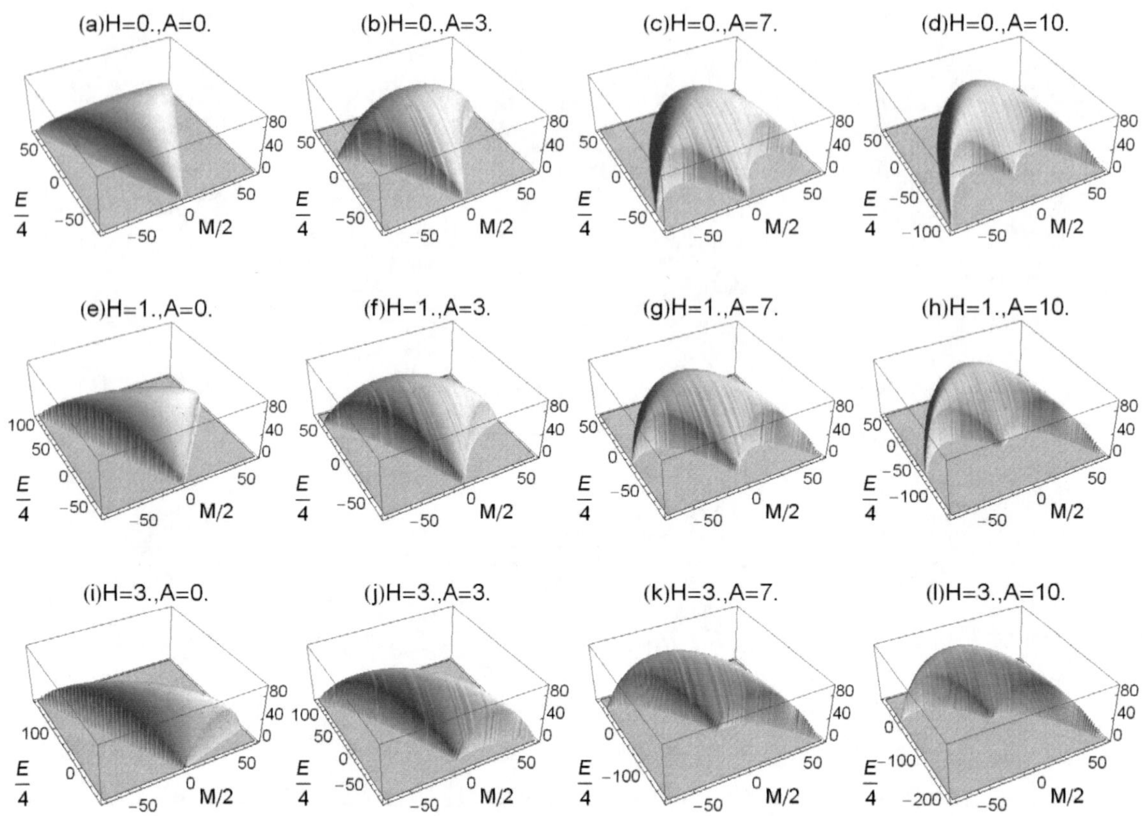


**Figure 2.** Plots of the logarithmic DOS vs energy ($E$) and magnetization ($M$), $\ln g(E, M)$, for different fields $H$ and long-range interactions $A$. The smaller system size, $L = 12$, is used for better graphical resolution. All possible values of $E$ and $M$ for this system size are considered.

negative $M$ are pulled up. When $A$ is applied, the energies for all states with non-zero $M$ are pulled down quadratically.

Figure 3 shows $g(E, M_s)$. Parts (a), (e), and (i) show the effect of increasing $H$ with $A = 0$. The energy range of states with $M_s$ near zero increases as states with negative $M$ are pulled up in energy, and states with positive $M$ are pulled down. The latter effect also causes the energy corresponding to the highest DOS to shift to a lower value. The other graphs in Fig. 3 show that increasing the long-range interaction strength $A$ lowers the minimum energy of states with $M_s$ near zero, as the energies of the corresponding states with large positive and negative $M$ are pulled down quadratically.

## 3. Summary

As an example of the application of the recently proposed macroscopically constrained Wang-Landau Monte Carlo method for systems with multiple order parameters, we have presented the joint DOS of a square-lattice Ising model with antiferromagnetic short-range interactions and ferromagnetic long-range interactions (a model spin-crossover material [8]). Based on data from simulation for one single parameter set, results are shown for various values of the model's field-like parameters, applied field $H$ and long-range interaction strength, $A$. Further results on

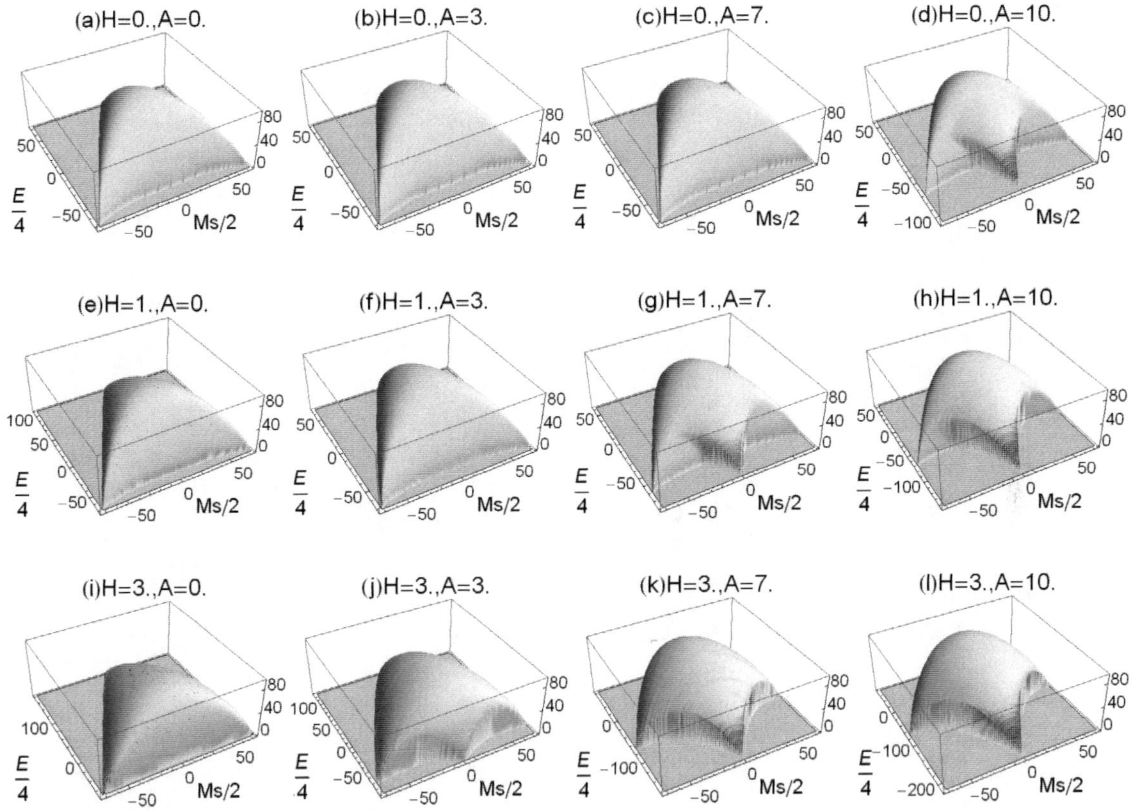


**Figure 3.** Plots of the logarithmic DOS vs energy ($E$) and staggered magnetization ($M_s$), $\ln g(E, M_s)$, for different fields $H$ and long-range interactions $A$, again using $L = 12$. All possible values of $E$ and $M_s$ for this system size are considered.

the phase diagrams of this model spin-crossover material are given in Ref. [5].

## Acknowledgments

Chan thanks Alexandra Valentim and Ying Wai Li for helpful discussions of the WL method. The Ising-ASFL model was first proposed by Seiji Miyashita, and we thank him for useful discussions. The simulations were performed at the Florida State University High Performance Computing Center. This work was supported in part by NSF grant No. DMR-1104829.

## References

[1] Wang F and Landau DP 2001 *Phys. Rev. Lett.* **86** 2050
[2] Wang F and Landau DP 2001 *Phys. Rev.* B **64** 056101
[3] Chan CH, Brown G and Rikvold PA 2017 *Phys. Rev.* E **95** 053302
[4] Ren Y, Eubank S and Nath M 2016 *Phys. Rev.* E **94** 042125
[5] Chan CH, Brown G and Rikvold PA 2017 *Phys. Rev.* B **96** in press. *Preprint* arXiv:1705.09465
[6] Nishino M and Miyashita S 2013 *Phys. Rev.* B **88** 014108
[7] Brown G, Rikvold PA and Miyashita S 2014 *Phys. Proc.* **57** 20
[8] Rikvold PA, Brown G, Miyashita S, Omand C and Nishino M 2016 *Phys. Rev.* B **93** 064109
[9] Waijnflasz J and Pick R 1971 *J. Phys. (Paris), Colloq.* **32** C1-91
[10] Lourenço BJ and Dickman R 2016 *J. Stat. Mech.* **2016** 033107

**Institute of Physics**
Dirac House, Temple Back
Bristol BS1 6BE UK

ISSN: 1742-6588
ISBN 978-1-5108-5341-6

CURRAN ASSOCIATES INC.
proceedings.com

# 5th Symposium on Prospects in the Physics of Discrete Symmetries 2016

Journal of Physics: Conference Series Volume 873

Warsaw, Poland
28 November - 3 December 2016

IOP
Institute of Physics